D1442235

BIRDS IN *Flight*

THE ART AND SCIENCE OF HOW BIRDS FLY

By Carrol L. Henderson

Voyageur Press

First published in 2008 by Voyageur Press, an imprint of
MBI Publishing Company, 400 First Avenue North, Suite 300,
Minneapolis, MN 55401 USA

Text and photography copyright © 2008 by Carrol L. Henderson
Illustrations copyright © 2008 by Steve Adams

All rights reserved. With the exception of quoting brief passages for
the purposes of review, no part of this publication may be reproduced
without prior written permission from the Publisher.

The information in this book is true and complete to the best of our
knowledge. All recommendations are made without any guarantee on
the part of the author or Publisher, who also disclaim any liability
incurred in connection with the use of this data or specific details.

We recognize, further, that some words, model names, and designations
mentioned herein are the property of the trademark holder. We use
them for identification purposes only. This is not an official publication.

Voyageur Press titles are also available at discounts in bulk quantity
for industrial or sales-promotional use. For details write to Special
Sales Manager at MBI Publishing Company, 400 First Avenue North,
Suite 300, Minneapolis, MN 55401 USA.

To find out more about our books,
join us online at www.voyageurpress.com.

Library of Congress Cataloging-in-Publication Data

Henderson, Carrol L.
  Birds in flight : the art and science of how birds fly / by Carrol L.
Henderson. — 1st ed.
      p. cm.
  Includes bibliographical references and index.
  ISBN 978-0-7603-3392-1 (hb w/ jkt)
  1. Birds—Flight. 2. Birds—Flight—Pictorial works. 3. Photography
of birds. I. Title.
  QL698.7.H46 2008
  598'.157—dc22
                                                              2008008070

Cover: Barn owl sequence. *Andy Harmer, WWI, Peter Arnold Inc*
Spine: Red-tailed hawk. *Carroll Henderson*
Front flap: Bald eagle. *FloridaStock, Shutterstock*
Title spread: Bald eagles. *FloridaStock, Shutterstock*
Frontispiece and back cover: Lanner falcon. *EcoPrint, Shutterstock*
Opposite and back cover: Ruby-throated hummingbird: *Ronnie
Howard, Shutterstock*
Contents: Tufted titmouse. *Steve Byland, Shutterstock*
Foreword: Goldfinches at feeder. *Tony Campbell, Shutterstock*

"Wandering Albatross," "Andean Condor," and "Epilogue" by
Pablo Neruda reprinted by permission of Fundación Pablo Neruda.
Copyright 1966 by Pablo Neruda. Reprinted from *Arte de Pájaros/Art
of Birds*, translated by Jack Schmitt, Lynx Edicions, 2002. For further
permission information, contact Fundación Pablo Neruda, Fernando
Marquez de la Plata 0192 Santiago de Chile, phone 56-2-777 87 41,
www.fundacionneruda.org.

Edited by Danielle Ibister
Designed by Cindy Samargia Laun

Printed in China

DEDICATION

This book is dedicated to distinguished ornithologist Dr. Harrison B. "Bud" Tordoff. Bud grew up in Mechanicville, New York. He received a B.S. from Cornell University in 1946 and a Ph.D. from the University of Michigan in 1950. During World War II, Bud served as a fighter pilot with the Eighth Air Force in Europe and became an ace after downing five enemy aircraft.

Bud served as a faculty member at the University of Kansas (1950–1957), the University of Michigan (1957–1970), and the University of Minnesota (1970 until his retirement in 1987). At the University of Minnesota, he served as the director of the Bell Museum of Natural History.

Bud is a lifelong admirer of the peregrine falcon because of its flying skills and beauty. He helped lead the Midwestern effort to save peregrine falcons. That effort, started at Cornell University, would restore a population of peregrines wiped out by DDT poisoning in the mid-twentieth century.

Since his retirement, Bud has continued to track and document the growth of the Midwestern population of peregrines. The status of the population has changed from no nesting pairs to a healthy and growing population of about 210 nesting pairs that raised more than 340 young in 2007.

The peregrine falcon is a special bird that embodies the ultimate avian qualities of speed, grace, and power in flight that Bud learned to appreciate as a fighter pilot long ago. The bird's successful restoration to the Midwest is a lasting legacy of Bud Tordoff's dedicated efforts.

Thanks, Bud.

Bud Tordoff

CONTENTS

FOREWORD

Since the earliest moment that we could call ourselves human, we have been mesmerized by the sight of birds in flight. The grace, the fluid movement, the effortless sweep and glide, the astonishing speed of a bird on the wing—these sights have inspired and challenged us for eons.

The dynamics of flight, which even a young bird on its first migration instinctively masters, still carry the aura of mystery. How can a hummingbird, wings a frenetic blur, hang motionless in the air—or fly backward? How can a ponderous condor get its bulk into the air? How can an albatross spend weeks on the wing, navigating the tumultuous winds of the sea, even in its sleep? How can a peregrine falcon come sizzling down from the high vaults of the sky, crackling through the air at more than two hundred miles per hour, descending on its prey like a guided missile?

Naturalist and wildlife biologist Carrol Henderson has spent his life watching birds with a photographer's eye and a biologist's training. Through his words and photographs, taken around the world, he leads us through the marvels of bird flight: the aerodynamics of life in the air, the adaptations of bone and feather and breath that make flight possible, and the sheer, unfettered beauty of birds in flight that still make us earthbound humans gasp in awe and envy.

Scott Weidensaul
January 2008

11

ACKNOWLEDGMENTS

I am indebted to many friends and colleagues for their assistance and encouragement while I was working on this very special project. My wife, Ethelle, provided continuing support as I researched and typed the manuscript and took the photos. Karen Johnson of Preferred Adventures, Ltd, deserves a special thank you, because many of the photos in this book were taken incidental to leading birding tours that she organized and coordinated. Dr. Noble Proctor and Dr. Scott Lanyon provided helpful scientific reviews of the text. Margaret Dexter, Jan Welsh, Judith LaFleur, and Lori Naumann also edited the manuscript. While most of the images in this book were taken of live birds in the wild, a few dead or mounted specimens or bird wings were used to illustrate specific aspects of wing structure and body profiles; the Nongame Wildlife Program in the Minnesota Department of Natural Resources provided a dead trumpeter swan and peregrine falcon for photography reference purposes. The staff of the Wildlife Rehabilitation Center of Minnesota was extremely helpful in making a selection of bird specimens available for photos, comparison, and study, including an indigo bunting and a common nighthawk. Jackie Fallon provided the photo of Dr. Bud Tordoff that is featured in the dedication. Taxidermist Peter Getman, who passed away in 2007, mounted the chimney swift that is featured in the book. He was the most skilled and accomplished songbird taxidermist I have ever known. Thanks, too, to the North Alabama Science Center in Huntsville for allowing the use of their retouched image of Leonardo da Vinci's ornithopter.

Common birds like mallards provide an excellent opportunity to view and photograph the phenomenon of flight in both rural and urban settings where they have grown tolerant of humans. *JD, Shutterstock*

INTRODUCTION

Visions of Flight

C hildren love to pretend that they can fly. Adults and children alike marvel at birds in flight. Even the most casual observer enjoys the beauty, symmetry, and grace of birds in flight. During a recent family visit, I was reminded of my own early interest in birds when my nephew borrowed the ornamental feathers of a peacock and rooster pheasant from my office so he could run across the backyard, flapping his arms and pretending to fly.

I grew up in central Iowa in the 1950s. Our farm near Zearing lay beneath a fascinating aerial crossroads. Geese migrated overhead, flying between summer nesting grounds in Canada and wintering grounds in the southern United States. With every sighting, my chores came to a halt as I listened to the honking geese and marveled at their mysterious V-shaped formations. Each memorable sighting created an intense curiosity about migratory birds, their annual cycles of life and survival, and most of all their ability to fly.

East-west flights above our farm created equal intrigue. Air Force jets from the Strategic Air Command passed high overhead on training missions from Offutt Air Force Base in Nebraska. Their white contrails created distinctive patterns in the sky. The jets provoked additional fascination with the marvel of flight.

What human has not, at one time or another, fantasized about having the ability to flap their arms and fly like an eagle? *Wesley Aston, Shutterstock*

The seasonal passage of wild geese never ceases to impress observers with the wonders of flight and the mysteries of migration.

Opposite: Pink flamingoes in Tanzania take flight from their tropical waterhole.
Sebastien Burel, Shutterstock

In the years that have passed since watching those geese and bombers fly over my boyhood home, I have experienced life as an officer in the United States Air Force and have flown in planes like those I watched as a boy. I have also become a professional wildlife biologist and an international birding tour leader. I have led more than 45 international birding tours and observed more than 2,500 species of bird in North and South America, Africa, China, Russia, and New Zealand.

I do not observe birds as a listing exercise. Birds should be watched and appreciated in the context of their habitats, behavior, adaptations, migrations, and unique flight characteristics such as hovering, gliding, soaring, flapping, and diving. Photographing birds during my travels has greatly enhanced my perceptions and understanding of bird life.

After twenty years of photographing birds, from hummingbirds to condors to albatrosses, I realized that many of the flight postures and behaviors captured in my photos demonstrated the broad spectrum of aerodynamic principles involving flight. They also captured an essence of the beauty in flight that requires no understanding of aerodynamics.

I draw equally on my background as a wildlife biologist and nature photographer in *Birds in Flight*. Whether you wish to learn the basic principles by which birds fly or you simply enjoy the beauty of birds in flight, the images in this book will remind you that wherever you live, birds can be a constant source of inspiration and enjoyment. I invite you to take the time to watch the birds around you and learn about the marvelous adaptations that make them such a wonderful part of our natural world.

PART I

The Art of Birds in Flight

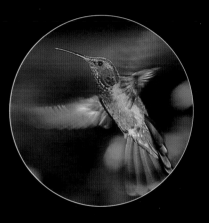

Above: When hummingbirds like this green violetear appear to float through the air and display their stunning iridescence, they can make a "birdwatcher" out of anyone.

Left: The activity of birding passes through several stages of development: discovering that there *are* birds, learning their names, making lists of birds seen, learning about ecology and survival, helping the birds, and finally embracing the art that birds, like this blue-and-yellow macaw, can inspire in our lives. *Shutterstock*

A *Gallery* OF BIRDS IN FLIGHT

While birdwatching, many people miss much of the wonder and beauty that birds have to offer because they place too much emphasis on simply seeing birds and making lists of the species they have seen. In the first section of this book, I attempt to share the "wonder" of birds in flight. The best aeronautical descriptions often fail to capture the beauty of flying birds. There is a point at which the appreciation of bird flight can be raised from science to the realm of art and poetry. This chapter is a photographic effort to capture that art by sharing some of the most enchanting and memorable moments I have experienced in my lifelong pursuit of birds.

Experiencing the art of birds in flight does not necessarily require travel to distant lands. You can discover this beauty in your own backyard, at local parks and nature centers, and at state and national wildlife refuges. If you take the time to watch the birds, the art and beauty of birds in flight will materialize before your eyes.

Sometimes the poetry of birds in flight can only be revealed through photographic methods that freeze the actions, as with this red-breasted nuthatch. *Jessica Archibald, Shutterstock*

Wandering albatross

A Poetic Portrayal of Flight

There is a fascinating and inevitable transition from appreciating the physical dimensions of avian flight to the realm of perceiving flight in art and poetry. Poet Pablo Neruda was an avid birder as well as a Nobel laureate. Many of his poems explore qualities of the birds he observed in his home country of Chile. The phenomenon of flight intrigued Neruda. In some of his poems, the image of a bird taking flight is a metaphor for a person reaching his or her full potential in life.

With only a few lines from the poem "Wandering Albatross" in his book *Art of Birds*, Neruda captures the essence of the wandering albatross:

> *The wind sails the open sea*
> *steered by the albatross*
> *that glides, falls, dances, climbs,*
> *hangs motionless in the fading light,*
> *touches the waves' towers,*
> *settles down in the disorderly element's*
> *seething mortar*
> *while the salt crowns it with laurels*
> *and the furious foam hisses,*
> *skims the waves*
> *with its great symphonic wings.*

Southern royal albatross

Like the wandering albatross, the southern royal albatross is master of the ocean skies. During its long life of forty to sixty years, an albatross may travel two million to three million miles over the wildness of southern oceans. This albatross was photographed near Kaikoura, New Zealand, as it soared past our boat. The passage of the albatross invoked a special silence among the eager birders. We knew we were in the presence of avian greatness. Such an encounter offers only a snapshot in time of a bird whose magnificent life circumscribes the southern seas.

White-capped albatross

This image brings back a magical moment when I was birding offshore Stewart Island, south of New Zealand's South Island. While watching albatrosses nearing our boat, I caught a movement from the corner of my eye and spun around just as this white-capped albatross passed within fifteen feet. I not only looked this awesome bird in the eye, I heard the wind rushing through its wings. I had just enough time to fire a snapshot at point-blank range. I will always remember this close-up meeting with a legendary master of the sea.

In addition to his fondness for the wandering albatross, Neruda also expressed his admiration for the Andean condor in *Art of Birds*:

> In the mountains the north wind
> whistles and howls like a missile
> and the condor leaves its casket,
> sharpens its talons on the rocks,
> spreads its mystical plumage,
> flies to the end of the sky.

This condor was photographed in southern Peru soaring through Colca Canyon, which is three times larger than the Grand Canyon. This famous birders' location allows visitors to see condors flying at eye level while exploring the panoramic landscape of the Andes Mountains. As our birding group stood on the canyon rim where the condor had passed, our response was again one of hushed silence. Spoken words would have spoiled the moment. There is something almost mystical about seeing an Andean condor. The bird inhabits some of the most rugged and spectacular terrain in the world, and it uses the mountain winds and canyons as its aerial highway to travel hundreds of miles with seemingly little effort.

Andean condor

Peruvian booby at Ballestas Arches

A Touch of Birds in the Landscape

Near the city of Paracas on the coast of Peru are some stunning rock formations known as the Ballestas Arches. While I was photographing the contrasting shadows and light of the arch, a Peruvian booby serendipitously passed through the arch. The touch of life and movement transforms the inanimate features of the rock formations and the crashing surf below to a scene of special beauty.

Visions of the Rainforest

My wife, Ethelle, and I have been leading rainforest birding trips since 1987. Every day in a rainforest is a new adventure because of the unpredictable appearances of birds and animals. During a two-week trip in Costa Rica, Ecuador, or Peru, it is possible to see three hundred to four hundred bird species.

The low light levels under the canopy presents a challenge when photographing rainforest birds. Some of the best birding occurs early or late in the day when the sun is low. Also, the photographer has little

Hoatzin

opportunity to shoot from a steady support or tripod. Most of the time photos are taken offhand from a boat or along footpaths in the forest. The philosophy of photographers in the rainforest is akin to the philosophy of grouse hunters in northern Minnesota: "Be ready and shoot quick."

One of my all-time favorite memories of bird-watching in the rainforest comes from a trip to south-eastern Venezuela. We were returning to our lodge in a dugout canoe on the Alambritos River. It was sunset. Primitive-looking hoatzins were perched in the riverside foliage and squawking as we passed. The hoatzin is a crested bird the size of a pheasant. One flushed and flew toward our boat. I had just enough time to raise my camera and find the hoatzin in my viewfinder. The light was so low that I almost didn't shoot, but I took the photo anyway as the bird passed by. Remarkably, the slow exposure captured the essence of the bird's flight and movement. I had moved the camera at the same speed as the bird was travelling and kept the eye in focus while the bird carried out a full wingbeat.

Blue-and-yellow macaw

Ringed kingfisher

One of my most suspenseful rainforest bird-watching experiences occurred along the Tambopata River in eastern Peru. Biologists at the Tambopata Rainforest Lodge have constructed a small blind on the edge of a cliff overlooking the river. Most mornings, hundreds of macaws, parrots, and parakeets swarm to the vertical cliff below the blind to eat clay.

On this particular morning, the birds began approaching the cliff a half hour before sunrise. The sun was still low in the sky when a squawking blue and yellow macaw flew in front of the blind. Even though the sky was still dark, I took the photo as the macaw flew by. The resulting picture was so dark that at first it was hard to see the macaw's image, but by brightening the picture, I found an abstract but beautiful macaw emerging from the predawn darkness—the essence of wildness in the Peruvian rainforest.

A great way to see tropical wildlife is to travel by river in a small boat. This traveling method allows birdwatchers to sight birds at close range. While exploring the canals of Tortuguero National Park in the Caribbean lowlands of Costa Rica, we flushed a ringed kingfisher from a low perch along the river bank. By panning my camera along the path of the bird as it rose, I kept the kingfisher in focus in front of the blurred riverbank foliage. When I examined the photo later, I discovered that the kingfisher was carrying a minnow in its bill.

The largest hummingbird in Costa Rica is the magnificent hummingbird. An inhabitant of high elevations, it is found around the Central Plateau and in the Talamanca Mountains south of San José to the Panama border. At the Savegre Mountain Lodge, the green hummingbird frequents the courtyard flowers. The curious magnificent hummingbird pictured here flew up and hovered, giving me just enough time to capture the essence of the hummingbird and its rapidly beating wings blending into the soft green light of its forested surroundings.

Magnificent hummingbird

The black vulture is the Rodney Dangerfield of the bird world. It gets no respect. Widespread throughout the southern United States and Latin America, the black vulture is a marvel of adaptation. It inhabits a variety of habitats and serves on Mother Nature's sanitation squad. It cleans up corpses of dead animals that would otherwise litter the landscape and allow diseases to pass to livestock or wildlife. Remarkably, physiological secrets in the stomach of the black vulture neutralize or kill deadly disease organisms and allow the vulture to consume the decaying carcasses. Most of the time, vultures are observed perched in trees or standing over some unfortunate dead creature. When spotted soaring over the countryside, however, birdwatchers can witness the grace of the vultures' movements as they scout the landscape for their next meal.

I observed this vulture from a hillside trail overlooking Rio de Janeiro in Brazil. We were birding as vultures circled and banked beside us. The birds were taking advantage of a slope thermal that rose up the mountainside and carried them past us at eye level.

Many of my photos from rainforests were taken in what would normally be considered less-than-ideal lighting conditions. However, I discovered that one way to convey the beauty and essence of flight is to capture the blurred motion and implied wing movement that is otherwise lost in a sharp photo.

Black vulture

BIRDS OF PREY

The augur buzzard shown here was flying across the savanna landscape of the Ngorongoro Crater in Tanzania. We were on a birding tour, and the buzzard flew in sync with our safari van. It stayed even with us for more than a hundred yards, allowing us to admire its strong and steady flight and to take some spontaneous photos. Perhaps the most impressive feature of this encounter was capturing the bird's distinctive rusty tail. This characteristic suggests that the augur buzzard is closely related to another hawk in the genus *Buteo* with which I am far more familiar in my home state of Minnesota: the red-tailed hawk.

Probably the most famous bird of prey in the world is the peregrine falcon. It will "ring up" to a great altitude—a thousand feet or more—to locate its prey. Once it has spotted a shorebird, jay, or teal, the peregrine folds its wings and takes a stunning dive. Flying at two hundred miles per hour, the bird covers hundreds of feet in a few seconds. With the skill of a fighter pilot, it continually adjusts its wings and tail as it closes in on its prey. The peregrine strikes with its closed talons, either killing or mortally wounding the smaller bird. The collision frequently results in an explosion of feathers. The peregrine then catches the prey in the air or follows it to the ground to capture, pluck, and eat its prize.

This female peregrine falcon was photographed atop the Wells Fargo Bank building in Bloomington, Minnesota. Biologist Jackie Fallon was placing a fluffy white chick back in the nest box after it had been weighed, measured, and banded. The female peregrine was ready to take us all on and part our hair in a way it had never been parted before. I took several quick photos and then retreated to safety. Banding is an essential part of the management and monitoring of Minnesota's falcon population. Biologists visit nesting sites only as necessary and only at times that do not place either the adult falcons or their chicks at risk. A hardhat is essential gear for such unusual photo outings.

Peregrine falcon

The crested caracara, also known as the Mexican eagle, is a bird of prey in the falcon family and is featured on Mexican currency as the country's national bird. It frequently preys on snakes, which endears it to locals. Its diet also includes small rodents, birds, lizards, invertebrates, and carrion. Caracaras are found from Texas south through Central and South America.

This crested caracara was landing on a perch in the Pantanal wetlands of Brazil. Sometimes, backlit photos convey the character of a bird more than a photo with standard front lighting. This striking silhouette captures the last view a mouse sees before it becomes lunch.

I photographed the American kestrel shown here from a car window near Riding Mountain National Park in southwestern Manitoba. The bird was perched on a power line. Just as I started to photograph, the bird dove into the grassy roadside and disappeared momentarily. With a mighty fluttering of wings, it emerged from the grass with a fat meadow vole. Struggling to gain flight, the kestrel dropped back into the grass and had to try again. The kestrel got a better grip on the still-struggling vole and flew off to a nearby field. As it disappeared from view, I realized that the job description of this small raptor includes "heavy lifting" as well as agile flight.

Above: Crested caracara

Opposite: American kestrel

Above: Pied oystercatchers

Below: Wrybills

Sky full of geese, Lac qui Parle

SKY FULL OF BIRDS

The wrybill is a rare and distinctive New Zealand shorebird. Its bill curves to the right, allowing it to reach under the edge of rocks and pebbles in search of small invertebrates. Like many shorebirds, wrybills fly in cloudlike flocks when moving among feeding and roosting areas. The huge flocks of pied oystercatchers and wrybills shown here were photographed at the Miranda Shorebird Center on the North Island of New Zealand. The sight of great flocks moving in such agile synchrony never ceases to amaze me. These birds serve as a reminder of how important it is to preserve wetlands for wildlife whether in New Zealand, Japan, Alaska, Minnesota, or your home state. As these flocks twisted and turned before me, a lone wrybill came hopping toward me along the beach. It was probing under pebbles for food. One of its legs was broken and useless, but it was surviving with one good leg and two good wings.

The sight of a sky full of birds has continually inspired me. In a world of declining species, it is refreshing to see success stories of birds that are still abundant or increasing in numbers. Until I worked at the Lac qui Parle Wildlife Management Area, I had never witnessed such magnificent avian gatherings. When the wildlife management area's manager Arlin "Andy" Anderson began his duties in 1956, he took on the responsibility of developing the area as a resting point for migrating Canada geese. He planted cornfields, placed decoys, and played recordings of goose calls. It worked. The first year, he attracted about 125 geese. By the time I was hired in 1974, the number of geese stopping each fall had increased to over 40,000. Now that number has increased to fall peaks of more than 100,000 geese. Their honking and V-shaped flocks command childlike awe as they fill the skies over the Lac qui Parle River valley. The sight and deafening sound of thousands of geese at sunset is a lasting tribute to the many conservationists, like Andy, who worked for decades to return these birds to our skies.

Swallow-tailed gull

TERNS AND GULLS

Terns are among the most elegant of all flying birds. They are known for hovering in the wind and scanning the water for prey. With the body raised to a vertical posture and wings uplifted to the point that the secondaries provide no lift, the primaries beat forward and backward to create just enough lift to allow them to hover motionless. While hovering, terns take on an angelic pose. This common tern was photographed above its nesting colony in Lake of the Woods, Minnesota. A recent storm had washed most nests in the colony into jumbles of speckled eggs at the high-water mark. The terns were returning to the sandy spit to begin renesting. The Forster's tern (page 38) was photographed at Oak Hammock Marsh Wildlife Management Area north of Winnipeg, Manitoba. It hovered fifteen feet in the air, looking for a minnow in the shallow water.

Exploring the Galápagos Islands is the adventure of a lifetime. The wildlife ignores the visitors in this pristine environment, allowing marvelous viewing. On South Plaza Island, a fifty-foot cliff creates an ideal slope updraft for swallow-tailed gulls and red-billed tropicbirds. They approach the cliff with ease to attend to their nests on the cliff face. The rare swallow-tailed gull only nests on the Galápagos and on another island near Colombia. It is nocturnal, a unique behavior among gulls. It flies at night to capture small squids and fish and to avoid harassment from frigatebirds, which are active during the day. The elegant markings of the swallow-tailed gull include red feet, a black head, and distinctive red eye-rings. The gull photographed here provides a study in contrasts. As it soars past, it conveys a sense of peace and tranquility while the whitecaps give evidence of a turbulent sea below.

Opposite: Common tern. *Carrol Henderson, Minnesota Department of Natural Resources*

Aerial Acrobats and Stunt Pilots

A fascinating outcome of wildlife photography is capturing a photo that reveals something totally unexpected. In this case, I was attempting to photograph a purple martin in flight when it flipped upside down like a stunt pilot. While upside down, it looked sideways to maintain its perspective with the horizon. It's no wonder that even fast-flying insects such as dragonflies and damselflies have little chance of escaping these acrobatic predators. This action happened so quickly that I didn't realize it had occurred until I examined my photos later. Even when we understand some of the physics of flight, birds can carry out subtle motions of flight at such high speeds that we fail to appreciate their complexity.

Opposite: Forster's tern

Above: Purple martin

Scaly-breasted hummingbird

During a visit to the Wilson Botanical Gardens in southern Costa Rica, I noticed a hummingbird regularly visiting the waxy, pendulous flowers of a Satyria plant. I set up my camera and flash on a tripod and focused on the bottom of the flowers in hopes of capturing the hummingbird pollinating the flower. I attached a twenty-five-foot remote cord to the camera and waited. I was not to be disappointed. The hummingbird came back twenty minutes later and flew to the flower. I pushed the button.

Not until I developed the film did I realize what I had captured. The scaly-breasted hummingbird was doing an avian backstroke. The photo depicts it hovering upside down in order to come up beneath the blossoms and insert its bill into the tube-shaped

Scintillant hummingbird

flowers—another miracle of flight captured by the camera that the human eye was unable to discern.

The tiny scintillant hummingbird is the smallest of Costa Rica's fifty-two species of hummingbird. Only 2-3/4 inches long, it lives at high elevations in the mountains of Costa Rica. One of my favorite locations for viewing this bird is Savegre Mountain Lodge, where it regularly visits the many flowers in the courtyard. This scintillant hummingbird, visiting Fuschia flowers, demonstrates its ability to hover with great precision as it feeds. Hummingbirds adjust the angle of both their body and wings as they hover, allowing them to perform remarkable maneuvers as they sip the nectar of small mountain flowers.

Snowcap

A rare bird eagerly sought by birdwatchers in Costa Rica is the diminutive snowcap. About three inches long, it is only found at a couple of highland lodges, including Rara Avis and Rancho Naturalista. The body flashes iridescent shades of deep royal purple, while the head is highlighted by its trademark snow-white cap. It creates an elfin vision that leaves the observer feeling enchanted.

This snowcap was visiting the "hummingbird meadow" at Rancho Naturalista. The staff provides several gallons of sugar water every day that attract hundreds of hummingbirds. The male snowcap hovered near me for just a moment as I quickly fired my camera. Some of my favorite hummingbird photos have been taken offhand with a flash as the bird approached or left a feeder or flower.

Ruby-throated hummingbird

This image of a ruby-throated hummingbird ranks as the biggest surprise I have ever had in attempting to capture the vision of birds in flight. I was on vacation with my family at Eagle Lake Lodge in central Minnesota and took some hummingbird photos near a feeder. I took dozens of photos in hopes that perhaps one or two would be in focus. When I got the film back from the processing lab a couple weeks later, I had to stare at this image for a while to figure out what I was looking at. The male hummingbird appears to be pinwheeling in the air and going three directions at once! The hummer had been approaching the feeder and suddenly decided to flee the area. The tail was still pointing at the feeder to the left, the wings were directed sideways, and the head was upside down. This avian contortionist was doing an extraordinary job of banking and turning all at once. This striking example reveals the humming-bird's extreme ability to maneuver, turn, and adjust its flight direction.

Egret Elegance

The snowy egret is an elegant wading bird. It never ceases to please the viewer with its immaculate plumage and graceful movements. A century ago, when women adorned their hats with the bird's lacey plumes, the snowy egret was nearly hunted to extinction in the United States. Now their populations have recovered, and they are abundant again.

This snowy egret was encountered along the canal that connects Tortuguero National Park to Moin in eastern Costa Rica. The flat lighting of the overcast sky subdued the brilliance of the white plumage and the white sky behind the bird. The egret flew silently past our boat and landed in a small tree on the bank ahead of us, where it turned and inspected us. After providing us with such a delightful sight, the egret began watching the birdwatchers.

When Thomas S. Roberts wrote *The Birds of Minnesota* in 1934, the great egret was unknown as a nesting species and rarely seen in the state. Since then, the bird has become a regular nesting species in southern Minnesota and its range has expanded northward. One of the egret's newest Minnesota colonies is on the east side of Fergus Falls on a small island in a city lake. The birds have become accustomed to visitors and offer one of the best viewing spots in Minnesota for great egrets.

Snowy egret

Great egret carrying stick

By waiting quietly in the park adjacent to the island, a person can witness egrets building nests, incubating their eggs, and raising their young. These sights are normally impossible to view, but the trees are short and the island is near the park road. The egret shown here was carrying a twig in its bill. It placed the twig in its nest atop one of the willow trees on the island. Such a sight reminds you of the various functions an egret's bill must undertake: catching fish, preening feathers, building nests, feeding young, and, if necessary, stabbing predators that approach the nest.

Above: Trumpeter swan, full flight

Below: Trumpeter swan, splashless landing

Trumpeter swan pair

MAJESTY OF THE SWANS

In 1982, I wrote a plan for the reintroduction of trumpeter swans as part of my work as Nongame Wildlife Program supervisor of the Minnesota Department of Natural Resources. Several years later, I began to fulfill my dream. This involved traveling to Alaska, collecting eggs in the wild, and bringing the eggs back to Minnesota for hatching and rearing. Two-year-old swans were released beginning in 1987.

When the released swans molted their clipped wing feathers, they could finally fly from the release site. Following the fall migration, swans return to the location where they learned to fly, so the trumpeter swans returned to the marshes of Minnesota.

I shot many photos during the trumpeter swan restoration project, but more than thirteen years after it began, I still had no good photos of a swan in flight. Meanwhile, the swans had established a wintering site along the Mississippi River near Monticello, Minnesota, where the hot water released from the power plant upstream keeps the river open throughout the winter. In January 1996, I visited Sheila Lawrence, who regularly fed the swans along the riverbank below her home. I learned that if I waited quietly, the swans would ignore me and fly right past me.

This photograph *(opposite, above)* both encapsulates the success of the trumpeter swan project and fulfills my dream of capturing the ultimate photo of a trumpeter swan in full flight. It conveys the power and grace of this great waterfowl species that can weigh up to thirty-five pounds and have a seven-foot wingspread. For more than one hundred years, this sight had been missing throughout much of Minnesota—a trumpeter swan on the wing. Now the swans are back, once again a part of the natural landscape.

When a swan comes in for a landing, it puts on an elegant performance. The wings arch in an angelic posture as its huge feet extend downward like an airplane lowering its landing gears. As the feet touch the water, momentum carries the bird forward as though it were waterskiing. This swan *(opposite, below)* made a splashless touchdown on the Mississippi River, scoring a perfect 10 for both style and technique.

Trumpeter swans are impressive for their size, their immaculate white color, and their graceful flight. They form lifelong pair bonds and have strong family ties. As they fly, they constantly vocalize to each other. Even more remarkable, their wings frequently beat in unison and create the vision of an elegant avian ballet.

Among all the photos that I have taken in the last twenty-five years, this one is my favorite. On a cold Saturday morning in January 1996, I went to the Mississippi River to see the trumpeter swans that I had worked to reintroduce to the state for over fourteen years. I was awed by the harsh and stunning scene on the river. The temperature had dropped to 25 degrees below 0 Fahrenheit during the night, and the river had nearly frozen over. Where the water was open, a constantly shifting fog formed over the water and drifted across the ice. The hardy trumpeter swans had gathered for the night in the last pool of open water.

Hoarfrost covered the tree branches overhanging the swans. The scene looked like something from a fairy tale. The swans barely moved and tucked their bills tightly under their wing feathers. I walked out on the ice to photograph the swans against the background of the crystalline forest. As I photographed, the sun rose above the riverbank behind them and shot a sunbeam through the frost-covered trees and onto the swans below. As the sunlight touched the swans, one of the magnificent birds rose upright and flapped its majestic wings.

It was a nearly angelic view of nature's beauty. I doubt I will ever have the privilege of seeing such a scene again. It was an impressive reminder that nature's beauty can be discovered at all times of year and in all types of weather conditions.

Trumpeter swans in the fog

Silhouettes

Sometimes the most striking photograph of a bird is not a portrait that showcases all of a bird's markings and colors but a backlit silhouette that captures a bird's profile against a sharply contrasting background. This black tern was passing over a Manitoba wetland when I pressed the shutter. The pointed bill was aimed forward and the sharply angled wings pointed back. The resulting profile conveys the essence of migration and calls to mind a compass needle marking its route. As these birds pass from northern nesting grounds to tropical and subtropical wintering areas, they use built-in navigation systems that rival the best global positioning navigation tools.

Hood Island is a "must see" island in the Galápagos Archipelago west of Ecuador. It is the home of Darwin's finches, Galápagos doves, blue-footed boobies, waved albatrosses, and the endemic Hood mockingbird. The best time to visit is early morning when the light is ideal for photography. This Galápagos hawk flew to the Hood Island geological marker just as we came ashore during our first visit to the Galápagos Islands in 1993. It paused to scan the landscape, then flew ahead over the low brush. Later, we found the hawk eating a Galápagos dove. One endangered species was eating another endangered species. In the Galápagos Islands, that is part of the balance of nature.

Carara National Park in Costa Rica is another nature lover's paradise. Below the bridge outside the park entrance, huge crocodiles sun themselves on sand bars along the banks of the Tarcoles River. The main path into the park provides an opportunity to see rainforest birds such as antbirds, trogons, and parrots. The featured species of the park, however, is the scarlet macaw.

If you visit the park at sunrise, you have the best chance to see scarlet macaws as you have never seen them before: wild and on the wing! They let you know

Black tern in the clouds

they are coming. Their harsh, guttural squawks can be heard over a half mile away—long before they are in view. Their squawking adds life and sound to the tropical rainforest. When they do appear on the horizon, they are almost always in pairs.

While standing on the bridge over the Tarcoles River in Costa Rica, I watched with excitement as a pair of scarlet macaws flew by and squawked in the predawn light. I took a picture of the macaws even though technically there was not enough light for a "decent" picture. When I later developed and enhanced the photo, I discovered the image of the two macaws against the soft background of a distant cloud—embodying the wild beauty of macaws flying over their predawn rainforest.

Scarlet macaw pair at sunrise

Galápagos hawk at sunrise

Black tern at sunset

SUNSET SPLENDOR

Black Rush Lake is a federal Waterfowl Production Area near Lynd, Minnesota. It has been restored by the U.S. Fish and Wildlife Service for the benefit of waterfowl and other wetland wildlife. An abandoned township road passes through the middle of the wetland and can be used as a walking path when the water is low. In July 2007, I discovered black terns nesting in the marsh. The terns were flying along the old road bed and searching for minnows in the water-filled ditches. Trying to photograph a rapidly flying tern is incredibly challenging. As I would focus on a tern near me, another one would often come in behind me and fly right over my head. It was actually a delightful challenge.

As sunset drew near, the distant horizon turned red over the expansive marsh. I spotted a black tern coming down the ditch channel toward me. It was the opportunity I had been looking for. As my autofocus locked onto the oncoming tern, I kept swinging the camera to track the tern as it passed by. As I pushed the shutter, my viewfinder filled with an explosion of red as the tern flew in front of the setting sun. When I checked my image later, I was delighted to discover that my black tern had become a red tern in the sunset. This experience was another reminder that some of the best wildlife photos are not planned. They just happen.

Gull-billed tern over the llanos

One of my favorite birding destinations in Venezuela is Hato El Cedral, a cattle ranch in the llanos region. The ranch has developed into a popular wildlife viewing destination for tourists. Encompassing more than 100,000 acres, the ranch features several 10,000-acre shallow impoundments that are flooded during the rainy season and managed as pastures to provide green grass for cattle during the annual dry season. Visitors can drive along the dikes of the impoundments or take boats into small channels for wonderful wildlife photography experiences.

The management system is good for the cattle and for birds. As surrounding wetlands dry up during the annual dry season, thousands of black-bellied whistling ducks, roseate spoonbills, scarlet ibises, jabirus, wood storks, egrets, herons, and terns flock to the ranch's mosaic of tropical wetlands.

As sunset nears and the birds retreat to their nocturnal resting places, the sky over the wetlands fills with an avian symphony of quacks, squawks, chirps, and honks. If you are lucky, you can find a good vantage point overlooking the marsh and watch as the red tropical sun balloons to a huge size just as it touches the horizon. During my first visit to Hato El Cedral in 1991, I experienced this marsh at sunset and took a photo just as a gull-billed tern tipped its wing into the setting sun. It was a beautiful scene made even more beautiful by the presence of a bird in flight.

PART II

Avian Aerodynamics

Above: The elegant great egret conveys a sense of immaculate natural beauty with its graceful flight.

Left: It is easy to watch a hummingbird and never question how it is able to float through the air. The study of avian aerodynamics allows a rewarding exploration into the miracle of flight. *Yanik Chauvin, Shutterstock*

2

Aerodynamic PRINCIPLES

Our lives are surrounded by birds in flight. Cities never have a shortage of pigeons. Ring-billed gulls circle parking lots and shopping centers looking for cast-off French fries. In the Twin Cities of Minnesota, I see Canada geese, mallards, red-tailed hawks, great egrets, great blue herons, double-crested cormorants, turkey vultures, bald eagles, and even ospreys on a regular basis. These sightings attest to the adaptability of birds to human environments and to the enduring phenomenon of flight itself. The phenomenon is even more fascinating when you understand some of the aerodynamic principles involved with avian flight.

The ability of a bald eagle to take off, dive, and soar is based on a little-known law of physics called Bernoulli's principle. *FloridaStock, Shutterstock*

THE MIRACLE OF LIFT

The principle of physics that lays the foundation of avian flight is known as Bernoulli's principle. Daniel Bernoulli was an eighteenth-century Swiss mathematician who recognized that air passes both under and over a bird's wing and meets at its trailing edge.

The wings of a bird are adapted for flight because of their aeronautical contours. The cross-section of a bird's wing is referred to as an airfoil. An airfoil is shaped like a long, slender comma on its side. It is thick and blunt at the leading edge, curves slightly upward in a convex pattern, and then tapers downward to a slender edge at the trailing portion.

The convex top of the wing presents a longer path for the air to pass over in the same amount of time as the air that is rushing along the flatter bottom surface of the wing. This discrepancy creates a difference in air pressure between the top and bottom surface. The top of the wing experiences less air pressure because the air is traveling at a higher speed to complete the longer path. This phenomenon generates lift that pushes the bird upward. Eventually, the speed of the air passing over the bird's wings becomes fast enough to generate a lifting force that equals the weight of the bird—and the bird becomes airborne.

Every bird has a minimum speed below which flight is technically impossible. The minimum flight speed for small birds is about eleven miles per hour, although most small songbirds fly at least sixteen to twenty miles per hour.

An interesting feature of Bernoulli's principle is that if the speed of a bird in flight doubles, the amount of lift quadruples. If the speed triples, the amount of lift generated multiplies nine times. This phenomenon explains why birds (as well as airline pilots) typically take off into the wind. The wind helps give extra lift. With any amount of wind present, the amount of lift produced increases significantly, in turn reducing the effort necessary to become airborne.

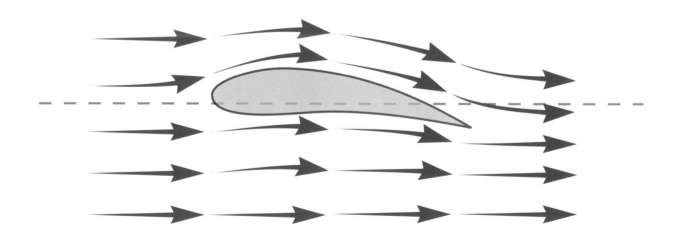

Bernoulli's Principle

The cross-section of a bird's wing forms an airfoil, in which the airstream over the top of the wing flows faster than the air under the wing. This dynamic creates less air pressure above the wing and allows lift for flight. When the force of lift equals the bird's weight, the bird becomes airborne.

Neotropic cormorants in Brazil take off into the wind. This helps them generate the lift they need to become airborne.

A SEA OF AIR AND WIND

An early breakthrough in understanding flight came when scientists Leonardo da Vinci and Sir Isaac Newton studied the properties of objects propelled through the air. They recognized that air was essentially an invisible fluid, and that birds fly through a sea of air. A bird's wingbeats, in a general way, simulate a swimmer's breaststroke or a pair of propellers moving through water.

Over millions of years of evolution, birds have become masters of the air. With each wingbeat and each aerial maneuver, they are constantly adjusting to prevailing winds, waves, updrafts, and the flight of nearby birds. Most of these adjustments are accomplished with such ease and speed that they go undetected by the human eye.

Flocks of birds on land or water often appear to be synchronized as they feed or swim. This alignment is because they are all facing into the wind. If flushed, they can take off into the wind.

Once a bird is airborne, it takes advantage of prevailing winds to reduce the energy needed for flight. If a duck is flying at thirty-five miles per hour, a tailwind of twenty miles per hour gives it a speed of fifty-five miles per hour.

When landing, birds typically face into the wind. This technique allows birds to better control their landing posture as they reach their stalling speed. It also helps them decelerate more efficiently before touchdown. Again, airplane pilots do the same.

A bird becomes airborne when the upward force of lift equals the downward weight of a bird (the force of gravity) and the forward thrust equals the drag created by the bird's body.

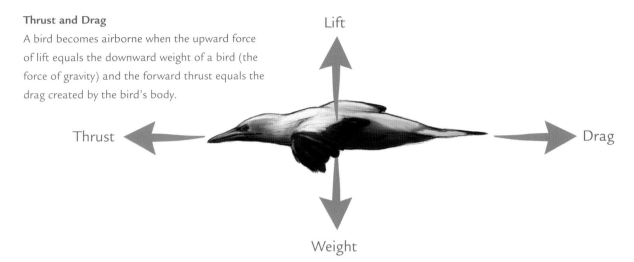

Lift

Thrust

Drag

Weight

THRUST AND DRAG

For birds to become airborne, they must have enough forward speed to generate lift. The wings must not only generate lift, they must also generate the forward thrust that keeps a bird flying forward fast enough to create and sustain lift. Thrust comes from the downward movement of the primary wing feathers. Thrust overcomes the force of drag, which is caused by the shape and contour of the bird and its wings. When the force of the thrust equals the resistance of drag, the bird can achieve flight.

The great frigatebird of the Galápagos cuts an impressive slender frontal profile that minimizes the amount of drag created by its body in flight.

The streamlined beauty of a bird in flight becomes apparent when it is viewed head-on. Frigatebirds, terns, and albatrosses cut a knifelike profile when viewed from the front with wings outspread. They literally slice their way through the air when gliding or soaring. Because of their shape, these birds create a smaller amount of profile drag than birds that are not as streamlined. Profile drag is the resistance created by the bird's body and wings from a frontal perspective.

When viewed from the side, bird profiles are variable. Terns and swallows have a slender profile, while pelicans, gannets, and boobies have a heavy-bodied design. Induced drag refers to the air resistance and turbulence created by the shape of a bird's wing, specifically the contour of the wingtips and the length of the bird's wings. Remember that there is higher air pressure on the bottom of a bird's wing and less air pressure on top because of Bernoulli's principle. Thus, air flows out along the bottom of the wing to the wingtip and curls up to the top of the wing surface. This airflow creates a swirling pattern of air currents called a vortex that results in turbulence and induced drag at the wingtips.

Different wing shapes create different amounts of induced drag. A long, narrow wing or a wing with notches between the primary feathers causes less induced drag. A short, rounded wingtip—such as that of a grouse, pheasant, or turkey—creates more pronounced induced drag. Rather than soaring or gliding, these birds typically fly with a rapid flapping pattern. This prevents the flight speed from being significantly decreased by the drag. Their flights are usually fairly short, too, so this method does not place them at a great disadvantage.

WINGS AS PROPELLERS

A bird's wings work, to a certain extent, like the propeller on the front of an airplane or on the top of a helicopter. An airplane propeller has a unique twist in its structure so that, as it turns, the leading edge catches the air and forces it backward along the sloping contour of the propeller. This movement forces, or propels, the airplane forward.

Now visualize a flying bird, viewed from the front, in which both wings are raised perhaps 20 degrees above horizontal. When the wings flap forward and down, the wing changes its profile. The primary feathers at the leading edge are stiffer and are driven downward with the wingbeat. The more posterior primary feathers have veins that are farther back from their leading edge, so they tend to be more flexible and twist upward as the wing is driven downward. All of the primary feathers (usually ten feathers) collectively assume a propeller's contour as the wing is flapped downward.

Each wing is thrust downward in a power stroke through an arc of perhaps 40 to 60 degrees. That is where the comparison with an airplane propeller stops. Unlike the airplane propeller that continues in a circle, each wing reverses course through an upstroke to the former upraised position. So each bird actually has a pair of reciprocating propellers.

The primary feathers generate forward thrust with each downstroke. The overall propeller design serves to drive the bird forward and provide the thrust necessary for flight.

The Forster's tern exemplifies streamlined efficiency. The bird creates minimal profile drag as it flies. Its pointed wingtips minimize the amount of induced drag, created by the difference in air pressure between the top and bottom of the wings.

The primary feathers of a snowy egret assume the curving contour of an airplane propeller on the downstroke. The stiffer leading edge of the wing drives downward against the air. The up-curved posterior portion of the primaries angles upward and facilitates airflow to the rear, creating forward thrust.

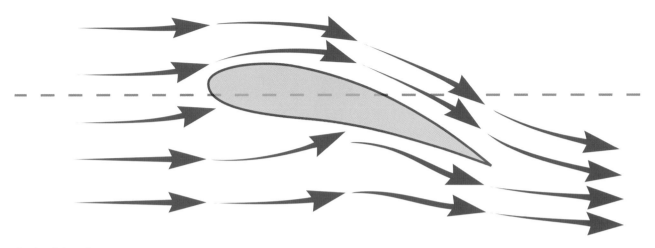

Angle of Attack

An airborne bird generates more lift by flying with its wings and body slightly upraised. This is called increasing the angle of attack. The air hitting the underside of the wings and belly generate an uplifting force in addition to the lift provided by Bernoulli's principle.

ANGLE OF ATTACK AND STALLING

The alignment of a bird's body and wings as it flies is called the angle of attack. If it flies straight ahead and parallel to the ground, the angle of attack is 0 degrees. When a bird slightly raises the front of its body, the airstream strikes the underbelly and the underside of the wings. The wind hitting the belly tends to push the bird upward. This provides additional lift without generating much additional drag. A bird in flight typically holds the front edge of the wing several degrees to about 13 degrees above horizontal to generate additional lift.

If the angle of attack is increased beyond about 16 degrees—such as when a bird is preparing to land—the airstream flowing over the top of the wing breaks away from the wing, creating a pocket of extreme swirling air turbulence on top of the wings and back.

At that point, the feathers over the back become ruffled and the bird stalls. If an airplane stalls, it falls from the sky. When a bird begins to stall, it has a special technique to defer falling until it is ready to land.

An Australasian gannet momentarily hovers by increasing its angle of attack to check out the area below for a potential landing site or the presence of fish.

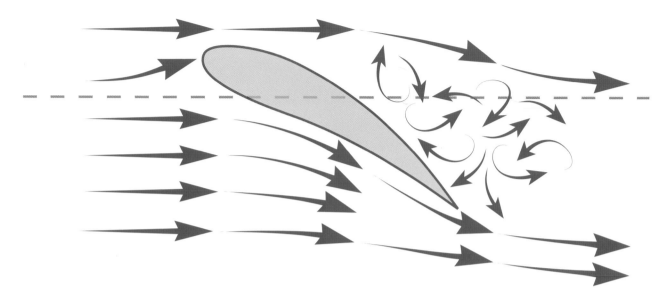

Stalling Angle

When a bird's angle of attack becomes acute, the airstream over the back is disrupted and the air becomes turbulent, ruffling the feathers over the back. This stalling normally occurs just prior to landing.

A snowy egret of Costa Rica lands with a steep angle of attack, causing the bird to stall and disrupting the airstream over the top of the wings. Turbulence ruffles the feathers over the back.

THE SECRET OF THE ALULA

There is a small bone on the leading edge of the wing at the wrist joint, comparable to the human thumb. It is called the alula. It serves as the anchor point for two to six short, stout, convex feathers. When a bird significantly increases its angle of attack prior to landing, these feathers are elevated and prevent premature stalling. The alula is explained in greater detail in Chapter 4.

Some birds can raise their angle of attack to nearly 70 or 80 degrees above horizontal by using their alulas and then going into a stall. At this point, the airstream over the back of the wings breaks loose and the feathers are pulled upward by the turbulence. Photos of great egrets and brown pelicans landing capture this moment of seeming levitation. With their large wings outspread like parachutes, they descend gracefully onto the tree branch of their choosing.

Aspect Ratio

Different bird species have different lifestyles. Bird wings have adapted in very different ways depending on how a certain species gets its food, where and how it builds its nests, and whether or not it migrates. One way to compare the differences in birds' wings is by using the aspect ratio, which is based on wing shape. Higher values of aspect ratio are associated with wings that are long and narrow, such as those of frigatebirds, pelicans, and albatrosses. Lower values are associated with wings that are short and wide.

The aspect ratio is calculated by dividing the square of a wing's length by the wing's surface area:

$$\frac{\text{Wing Length} \times \text{Wing Length}}{\text{Wing Surface Area}} = \text{Aspect Ratio}$$

For example, if a pelican's wing is 30 inches long and has a surface area of 90 square inches, the aspect ratio would be 10, based on the following equation:

$$\frac{30 \times 30}{90} = 10$$

The values for aspect ratios range from 15.0 for albatrosses to about 4.8 for small finches. Some examples of aspect ratio values are: wandering albatross, 15.0; magnificent frigatebird, 14.7; common tern, 13.2; brown pelican, 10.6; osprey, 8.9; American white pelican, 8.5; barn swallow, 8.0; Andean condor, 7.5; and ring-necked pheasant, 5.5.

Albatrosses have the highest aspect ratios among all birds. They have extremely long, narrow wings adapted for soaring at sea in sustained high winds. The long wings provide excellent lift and allow for long-distance travel. They do not allow great maneuverability.

The high aspect ratio of a pelican's wing allows the bird to spend much of its time soaring. Pelican wings do not provide for great speed or agility, but those qualities are not as essential for their survival or foraging needs.

Brown pelican: 10.6 aspect ratio

Ring-necked pheasant: 5.5 aspect ratio

The albatross has the highest aspect ratio of all bird species. Its long narrow wings are adapted for soaring at sea in high winds.

Opposite: The short, broad wings of the ring-necked pheasant are adapted for explosive flushing from the ground to surprise potential predators and facilitate escape. Its wings are not adapted for long flights. *Taxidermy by Carrol Henderson*

The aspect ratio of a bird's wing is not a fixed value. This composite photo shows two views of a Franklin's gull in flight. When the gull folds its primary feathers backward (*right*), the aspect ratio and amount of drag decreases. The center of gravity shifts back and the gull accelerates.

The ring-necked pheasant has a wing aspect ratio of 5.5. The short, rounded profile of the wings, together with the long tail, allows for explosive flushes, great maneuverability, and relatively short flights. The pheasant is not migratory and its flights typically last less than a mile. Its wings also make a noisy, explosive whirring. These characteristics allow the pheasant to escape quickly from both avian and terrestrial predators in wooded and open habitat. The long tail may also divert the attention of a predator, who bites at the tail as the bird flushes, leaving it with only a mouthful of feathers.

Characteristics such as aspect ratio suggest constant values, but in fact the shape and aerodynamic qualities of the wings change from moment to moment. As a bird flies, the wing profile can be changed to facilitate speeding up, slowing down, or diving. When a wing is folded back, the surface area of the wing and the aspect ratio decrease. When a wing is extended outward, the surface area and the aspect ratio increase.

Two views of a Franklin's gull are shown here. The left image shows the gull in typical flight, with outspread wings generating maximum lift. When the primary feathers are folded backward, as shown on the right, the aspect ratio decreases. The folded wings offer less profile drag, and the gull increases its airspeed. This technique works well when the gull is gliding. As the gull adjusts its wings in this manner, the center of gravity shifts farther back and the front of the bird inclines slightly downward. This facilitates diving and accelerating toward prey. This principle is also used by diving boobies and falcons to increase airspeed as they dive toward prey.

The ability to change wing configuration, like the Franklin's gull, has been a long-standing goal of military aircraft designers. The first such aircraft was the experimental Bell X-5 plane in 1951, followed by development of the F-111 fighter bomber in 1967, the F-14 Navy Tomcat in 1976, and the B-1 Lancer strategic bomber in 1986. The F-111 had a wingspan of 63 feet when the wings were fully extended. That configuration offered maximum lift for takeoff and landing and low-speed maneuvers. The wings folded back to 32 feet, reducing drag and increasing maximum speed. With folded wings, the F-111 could fly at two and a half times the speed of sound at altitudes up to 60,000 feet. The F-111 had an aspect ratio of 7.56 with the wings fully extended and 1.95 with the wings swept back. The B-1 Lancer is the only variable sweep plane still in the inventory of the U.S. military.

WING LOADING

Wing loading is the ratio of the bird's weight to the wing surface area. It measures how much wing area is available to provide the lift necessary to support a bird's weight in flight.

Wing loading is calculated by dividing the weight of a bird by the total surface area of both wings:

$$\frac{\text{WING AREA}}{\text{BIRD WEIGHT}} = \text{WING LOADING}$$

The diagram illustrates a range of metric measurements from high to low wing loading values for four birds: 0.56 cm²/gm for the common loon, 1.81 cm²/gm for the turkey vulture, 3.40 cm²/gm for the brown pelican, and 6.96 cm²/gm for the barn swallow.

High wing loading is characteristic of heavy birds with relatively small wings. They are typically waterbirds that require an extended runway of open water to attain enough speed to take off, such as loons, swans, geese, cormorants, diving ducks, mergansers, and grebes. These waterbirds take flight by facing into the wind, running across the water, and vigorously flapping their wings. They have difficulty taking off in calm weather. They must beat their wings fast to generate the forward speed necessary to produce the lift force that equals their body weight.

In calm weather, a common loon may need to run several hundred feet on the water to gain flight. On a windy day, a loon can take off into the wind over a much shorter distance. The loon begins by facing into the wind and flapping its wings hard as it paddles with its feet. Then it runs across the water's surface with wingtips slapping the water. Each stride of the huge webbed feet kicks up an artistic spray of water droplets. Finally, the loon reaches enough speed to become airborne.

1. cm²/gm = 6.96
Barn swallow

2. cm²/gm = 3.40
Brown pelican

3. cm²/gm = 1.81
Turkey vulture

4. cm²/gm = 0.56
Common loon

Wing Loading

This diagram shows four levels of wing loading. The barn swallow has low wing loading. It has a large wing surface to support its body weight and a wing loading value of 6.96 square centimeters of wing area for each gram of body weight. Intermediate values are shown for the brown pelican and turkey vulture. The common loon represents high wing loading.

The trumpeter swan is the heaviest waterfowl species in the world. An adult may weigh thirty-five pounds and have a wingspan of seven feet. To take off, the swan slaps the water with its wings as it heads into the wind and begins paddling and then running on the water. It may run several hundred feet across the water to achieve flight. Once airborne, the swan's great wingspan makes it a master of the sky.

In each case, the waterbird needs to attain enough speed to create the force of lift that will equal the bird's weight in accordance with Bernoulli's principle. For birds with high wing loading, this requires a significant—often extended—takeoff effort.

Low and medium wing loading is characteristic of birds that have large wing surface areas in relation to their body weight. These birds include frigatebirds, pelicans, hawks, eagles, owls, swifts, hummingbirds, swallows, terns, kestrels, herons, and egrets. Low wing loading works to the advantage of birds that hunt from the sky. They need to fly slowly enough to scan for prey.

The albatross is an unusual example of a bird that spends most of its life soaring but has high wing loading. It lives in an extremely windy environment that provides the additional lift needed to soar for extended periods.

The turkey vulture is another widely known soaring bird that does not have low wing loading. It compensates by taking advantage of thermals and updrafts for soaring as it searches for carrion. Riding warm air currents makes up for its high wing loading value.

With a wing loading value of 1.81 square centimeters of wing area for each gram of body weight, the turkey vulture does not have a very low wing loading value for a bird that spends so much time soaring in search of carrion. However, it compensates for that potential shortcoming by soaring on thermals and updrafts.

Wingtip Vortex

Higher air pressure under a bird's wing, caused by Bernoulli's principle, tends to flow past the wingtip and then spiral backward. The updrafting portion of this vortex of air is used by trailing birds in a V-shaped formation to achieve extra lift.

WINGTIP VORTEX

The wingtips of birds provide some critical benefits for flight. Because of Bernoulli's principle, uneven air pressure is applied to the bottom and top wing surfaces. The higher pressure below flows from the bottom of the wing to the top of the wing at the wingtip, instead of along the trailing edge of the wing where the constant airflow prevents such flow from bottom to top. When air flows around the wingtip of a bird such as a goose or pelican, it swirls upward in a circular pattern, creating a spiraling current of air swirling back from each wingtip. This effect is referred to as the wingtip vortex. As this vortex swirls upward just outside and to the rear of each wingtip, it creates a pocket of updraft currents for several feet behind each wingtip. The amount of wingtip vortex created varies depending on the type of wingtip. Pointed and slotted wingtips create less air disturbance than blunt, rounded wingtips.

An eastern chanting goshawk *(above)* of Kenya and a sandwich tern *(opposite)* of Costa Rica demonstrate the venetian blind effect. The primary feathers separate and twist upside down as the wings flip backward and upward in preparation for the next forward power stroke.

Venetian Blind Effect

As a tern or kestrel hovers overhead, as a Canada goose descends to the water, and even as a mallard drops into a field of ducks, an intriguing phenomenon occurs with the upstroke of each wing. The "arm" portion of the wing (humerus) is drawn up and close to the body. The "forearm" section (radius and ulna), with the secondary feathers, is also directed upward. The "hand" sections (primary feathers) rapidly beat forward and then turn upside down and flip backward. The wings function momentarily as reciprocating helicopter blades, allowing extended hovering for terns and kestrels and momentary hovering for heavier waterfowl.

As the wings carry out their upside-down backstroke, the primary feathers twist in their sockets and separate from each other like the panes of a venetian blind. This motion allows air to flow rapidly between the feathers. The upside-down primaries are held at a slightly upraised angle, creating an increased angle of attack for each of the primary feathers and generating a small amount of lift during the backstroke. The separated feathers also cause much less drag.

This phenomenon could not be perceived by early students of flight such as Leonardo da Vinci. He thought flight was achieved by simple up-and-down flapping of the wings. Only with the aid of high-speed photography in the last century could the aerodynamic functions of the wing upstroke and the venetian blind effect be interpreted and understood.

3

Feathers AND Bones

Feathers and bones function as the building blocks of bird flight. Unique to birds, feathers provide adaptations for both survival and flight. Bird bones are uniquely adapted to provide strength without adding substantially to the bird's weight. The frigatebird is so specialized for soaring that even though it has a seven-foot wingspan, its feathers actually weigh more than its ultralight bones. The bones of a frigatebird weigh only four ounces!

Owl feathers are adapted for silent flight so that prey will not hear the owl approaching.
Dariush M., Shutterstock

FEATHERS

Feathers insulate, camouflage, attract mates, intimidate intruders, and of course adapt birds for flight. The modified wing feathers of a snipe can make sounds in flight, the waterlogged breast and belly feathers of a rhea can help transport moisture to overheated eggs, and other specialized feathers help with tactile senses, swimming, and hearing. Feathers are strong, durable, flexible, and light. All of these qualities are essential in order for birds to fly.

The kinds of feathers necessary for flight include primary wing feathers, secondary wing feathers, and scapular feathers. Contour feathers are outer feathers on the body that provide a smooth and aerodynamically streamlined profile. Down feathers provide extremely lightweight insulation. Tail feathers serve as a rudder for a bird to adjust the direction of its flight upward, downward, or sideways.

Feather oiling and maintenance. Since feathers need to be flight-worthy for about a year, they require constant maintenance to retain their waterproof qualities, to clean off dirt, and to align ruffled feathers. Some birds spend up to 30 percent of their time preening. Preening involves cleaning and oiling the feathers. Most birds have a uropygial gland on their back, just in front of the tail. This gland produces oil that the bird squeezes out with its bill. The bird then strokes its feathers with its partially closed bill to apply this waterproofing oil. Preening also helps strip parasitic feather mites from the plumage.

Instead of uropygial glands, herons and bitterns have patches of feathers on their back called "powder down." These feathers, which the bird strips off with the bill, crumble into a fine talcum-like powder that provides waterproofing benefits.

Weatherproof feathers. Feathers must function in temperature extremes: from the dry heat of deserts to the icy cold of a winter tundra. In some avian environments, temperatures drop lower than 30 degrees below 0 Fahrenheit. Feathers must also be waterproof and not become covered with ice during subfreezing weather.

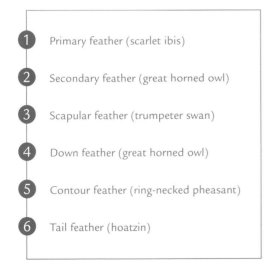

1. Primary feather (scarlet ibis)
2. Secondary feather (great horned owl)
3. Scapular feather (trumpeter swan)
4. Down feather (great horned owl)
5. Contour feather (ring-necked pheasant)
6. Tail feather (hoatzin)

A scarlet ibis squeezes oil from the uropygial gland above its tail. It applies the waterproofing oil by pulling individual feathers through its bill.

Feathers and bones are the building blocks of flight.

Feather replacement. Because of wear and tear, bathing, preening, fighting, and the effects of external parasites, feathers eventually become worn. The process by which feathers are systematically replaced is called molting.

Ducks, geese, and swans typically molt once a year, after the nesting season. In the United States, many waterfowl lose their feathers in June, July, or August. They pass several weeks in a flightless condition, making them vulnerable to predators. However, because waterfowl nest in wetlands, they can hide out from predators in the dense reeds and foliage of their habitat. They continue to feed on aquatic vegetation and invertebrates while flightless, and new feathers grow in time for fall migration.

Most other birds cannot afford to lose their flight ability. The kelp gull, for example, must fly every day to locations perhaps miles from the colony to catch the fish it needs to survive. It loses its feathers a few at a time in a systematic progression of feather replacement. Many songbirds also molt their feathers a few at a time.

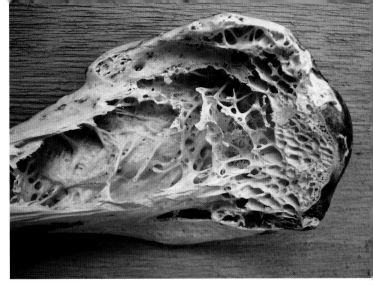

Fine struts within the wing bones, as revealed in this cross-section of an American white pelican humerus, provide great structural strength without adding much weight.

Trumpeter swan feathers exhibit extreme durability, retaining their aerodynamic qualities for flight even when temperatures reach 30 degrees below zero.

BONES

Birds typically have lightweight wing bones with inner reinforcements that provide exceptional strength. Within the wing bones of pelicans lie a network of "struts" that add structural strength to the thin walls of the bones. These struts function similar to the struts used in biplanes to provide strength between the upper and lower wings. The bones of a brown pelican weigh only about twenty-three ounces while the bird may weigh over seven or eight pounds.

The structure in a bird's wing bones closely follows the structure of a human arm. The wing bone closest to the body is the humerus. The next two paired bones are the radius and ulna. The ulna, on the trailing edge of the wing, has secondary feathers attached to the back edge of the bone. At the joint where the radius and ulna join to the "hand" bones, there is an interesting deviation from the structure of the human hand. The "thumb" on a bird is a specialized bone called the alula that lies along the leading edge of the wing. Beyond the alula are the metacarpal bones and two extremely long, slender fused bones that correspond to the human index finger and middle finger.

A kelp gull of New Zealand displays multicolored wing feathers. The faded feathers are older; the black wing feathers have just grown back after molting.

Wings

The wings of a bird tell a wonderful story, not only about flight and aerodynamics, but about the natural history of the bird itself. Wings have developed for each species through natural selection to provide the ideal qualities needed to survive in a particular habitat, find mates, migrate, fight with competitors, and seek out food. From the tiny wings of a hummingbird to the five-foot wings of a soaring albatross, every wing tells a story of how a bird has adapted to its environment.

Perhaps the most amazing thing about bird wings is that, unlike airplane wings, they continuously change shape. The changeability allows a bird to speed up, slow down, and even pause momentarily to hover while examining the ground below for prey. If a peregrine falcon has its wings completely outspread for soaring, it generates more lift. While soaring, the primary feathers are held straight out. If the bird spots prey below, it will "ring up" in great circles until it reaches an elevation hundreds of feet high, where, from a birdwatcher's perspective, the falcon is only a speck in the sky.

Among the waterbirds that can be enjoyed in the Midwestern wetlands, the Forster's tern embodies the essence of flight with its elegant form and graceful movements.

After the peregrine targets its prey, it banks and begins to dive. The bird partially folds back its wings, reducing profile drag and allowing it to accelerate. With the wings folded back, the center of gravity moves backward and the falcon is inclined downward. The tail also cocks downward, to help adjust to the downward angle of flight.

As the falcon dives, the primaries are progressively folded straight back and held tighter to the body. This further reduces the drag. The bird shifts the center of gravity farther back and increases airspeed. Minor flicks and twists of the primaries and tail provide thrust and adjustments to the trajectory as the plummeting falcon tracks its prey like a guided missile. The peregrine becomes a feathered bullet dropping out of the sky at nearly two hundred miles per hour. With its talons clenched into fists, the falcon strikes its prey, resulting in an explosion of feathers. Falconers consider the peregrine the ultimate bird because of its marvelous adaptations for extreme flight.

PARTS OF A WING

Wing feathers, from the body to the wingtip, include four types: scapular feathers, secondary feathers, primary feathers, and alula feathers.

Scapular Feathers

These feathers overlay the wing feathers on the back when the wing is folded to the body. They provide a streamlined transition in the aerodynamic contour of the bird between body and wings.

This composite photo shows several postures of a peregrine falcon diving for prey. As the wings are drawn in and folded back, the drag decreases and the airspeed increases to nearly 200 miles per hour.

Secondary Feathers

The trailing feathers associated with the radius and ulna are called secondaries. The cross-section of this portion of the wing creates the airfoil that provides lift for a bird in flight. The number of secondary feathers varies among different species. Hummingbirds have only six secondary feathers; both lift and thrust are provided by the primaries while hovering. Birds that depend on soaring for their lifestyle have a higher number of secondaries.

Most songbirds and other small birds have medium-length wings and nine to eleven secondary feathers. Great horned owls have fourteen secondaries. Turkey vultures have longer wings adapted for soaring and have eighteen secondaries. Ospreys have twenty. Andean condors and bateleur eagles have twenty-five; both species are adapted for extended soaring and gliding as they search for carrion in the mountains of South America and on the plains of Africa, respectively. Wandering albatrosses have thirty-two secondary feathers on their long, narrow wings. These long rows of secondaries create considerable lift while minimizing the amount of energy necessary to stay aloft for extended periods.

For many larger birds such as cranes, geese, and swans, the section of the wings with the secondaries is held in a relatively horizontal position to create lift when the primaries flap up and down to generate thrust. The efficiency of the secondaries in creating lift is higher than if that portion of the wing were also flapping in a larger arc.

Primary Feathers

The distal section of the wing holds the primary feathers. Primary feathers are attached to two fused bones that correspond to the index and middle fingers of a human hand. Birds have nine to twelve primary feathers attached to the back edge of these fused "finger bones." They provide thrust for moving through the air during flapping or hovering flight.

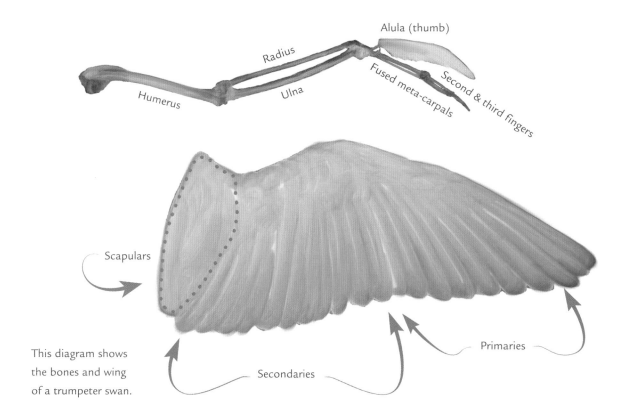

Alula (thumb)

Radius

Humerus

Ulna

Fused meta-carpals

Second & third fingers

Scapulars

This diagram shows
the bones and wing
of a trumpeter swan.

Secondaries

Primaries

Birds that soar have specialized primary feathers. The leading four to six primaries are wide at the base and narrow at the tip, creating a slotted effect that allows air to pass through and reduces drag at the wingtips. The slots reduce the turbulence that would otherwise occur from the wingtip vortex effect of blunt-ended wings. This specialization creates a smoother airflow in which each slotted primary feather reduces the induced drag at the wingtip while flapping and soaring, in much the same way that the alula reduces turbulence over the top of a wing when a bird is landing. The cross-section of each individual primary feather also generates its own individual Bernoulli's principle and provides some additional lift at the wingtips. This is why the primaries of birds with slotted wingtips are often upturned while in flight.

Alula Feathers

The fourth section of the wing is part of the leading edge on top of the wrist joint. It is called the alula and corresponds to the thumb on a human hand. Two to six stiff, concave feathers are attached to the alula. The alula comes into play when a bird lands at an angle of attack exceeding 16 degrees. When the alula is raised, it becomes a "slat" that forces an intense airstream along the top surface of the wing. This prevents stalling as the bird's forward speed decreases and as the angle of attack increases even more. Finally, the angle of attack increases to the point that the bird intentionally stalls and the airstream over the top of the wings breaks away from the surface of the wings and the back. At that point, the birds uses its outspread feet, tail, and wings to slow as it descends to its perch.

The lift that allows the brown pelican to soar so effortlessly is generated by the "arm" sections of its wings that include the secondary feathers.

The constricted area at the tip of a red-tailed hawk primary feather creates a slot on the wing. The slots create an airstream through the wingtips that reduces turbulence and drag.

An interesting piece of aircraft history is associated with the function of the alula. In 1919, British aircraft designer and researcher Frederick Handley Page patented a design to create a slat device along the upper leading edge of an airplane's wings that had the cross-section of an airfoil. The device could be extended forward during landing and force airflow between the slat and the leading edge of the wing. This design reduced turbulence along the top surface of the airplane wing, provided more lift during landing, and reduced the chance of stalling and crashing the airplane. In essence, Handley Page had reinvented the avian alula for use on an airplane without knowledge of the comparable structure on birds.

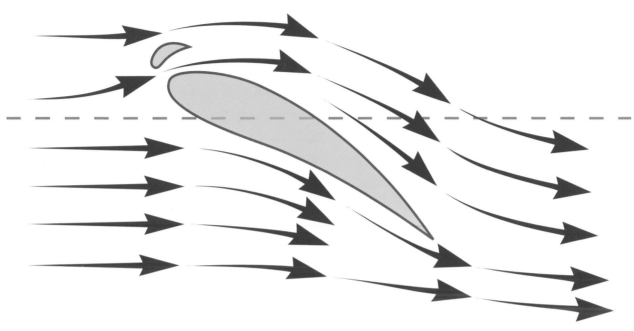

Alula and Airflow

The upraised alula forces air over the top of a bird's wing and prevents stalling. This characteristic allows the bird to remain airborne at low speed prior to landing. The Handley Page slat performed the same function on early aircraft wing designs.

TYPES OF WINGS

Traditional ornithology texts describe four types of bird wings: 1) long, narrow, pointed soaring wings like those of albatrosses; 2) long, broad, soaring and flapping wings with slotted wingtips like those of hawks, vultures, eagles, and condors; 3) long, angular, pointed, high-speed wings like those of falcons, terns, swifts, and hummingbirds; and 4) short, broad, rounded wings like those of grouse and songbirds.

However, I distinguish between six types of wings, based on distinctions in flight characteristics. The short, rounded wings of grouse are used quite differently than those of songbirds, so I have described them in separate sections. Also, the high-speed wings of falcons, terns, and swifts are used quite differently than hummingbirds so they are also described in separate sections.

A descending Canada goose extends its alulas above each wing. Each alula directs an airstream over the wing and prevents stalling at low airspeed.

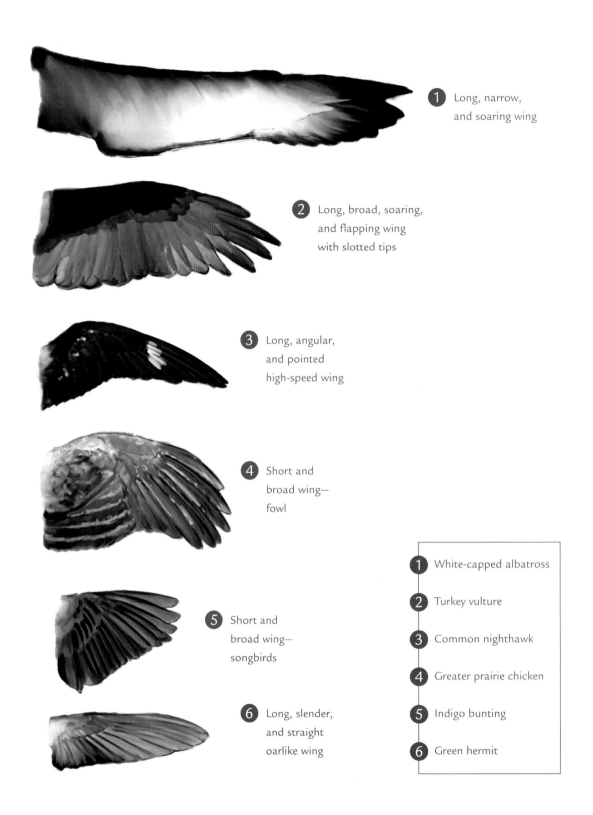

1 Long, narrow, and soaring wing

2 Long, broad, soaring, and flapping wing with slotted tips

3 Long, angular, and pointed high-speed wing

4 Short and broad wing— fowl

5 Short and broad wing— songbirds

6 Long, slender, and straight oarlike wing

1 White-capped albatross

2 Turkey vulture

3 Common nighthawk

4 Greater prairie chicken

5 Indigo bunting

6 Green hermit

The white-capped albatross has extremely long, narrow wings that give it exceptional lift where it searches for food at the ocean's surface in its windy marine environment.

The long, narrow wings of the osprey are adapted for migratory flights and for agile aerial maneuvers as it dives for fish in marine and freshwater habitats.

1. Long, Narrow, Pointed Soaring Wings

The albatross has long, narrow, pointed wings adapted for flight in windy oceanic environments. The constant wind provides its heavy body with enough lift to glide above the waves searching for food for thousands of miles. Pointed wingtips minimize air turbulence and diminish induced drag. This increases the efficiency of its soaring flight.

The albatross has a relatively heavy body considering the amount of wing area that generates lift. Its high wing loading would suggest that the large bird must expend a significant amount of energy in flight. However, the high wind speeds that characterize its environment compensate for the small surface area of its wings. The wind helps it maintain soaring flight for long distances with minimal energy expenditure. If the wind dies, the albatross has great difficulty taking off from water and may be stranded for days. It also has the highest aspect ratio of all birds. This is an advantage for soaring and gliding in the wind because the long wings generate extra lift.

2. Long, Broad, Soaring, and Flapping Wings with Slotted Wingtips

Pelicans, vultures, condors, hawks, ospreys, ibises, flamingoes, geese, and swans all have long, broad wings adapted for flapping, and some are also adapted for extended soaring. Their wings provide great lift. They are characterized by a medium to high aspect ratio and low to medium wing loading. Most of these birds have slotted wingtips. This refers to the conspicuous separation of the outermost primary feathers, which reduces air turbulence at the wingtips and provides some additional lift while soaring and gliding.

Ospreys have long wings that provide considerable lift for long flights in search of fish and migratory flights. The slotted wings have long primary feathers that generate significant thrust both in flight and while diving for prey. The primary feathers can also be upraised to allow the osprey to "wind hover" and remain motionless as it scouts for fish in the waters below.

Notice the dihedral posture of the turkey vulture's wings. The wind has caused the vulture to tip to the right. When this happens, the horizontal position of the right wing generates more lift than the left wing. The bird's "self-righting" posture raises the right wing back to the original dihedral position.

Dihedral Wings

Among the most conspicuous of the soaring birds in the Americas is the ubiquitous turkey vulture. This species can be found from Canada to Costa Rica and in many countries of South America. The turkey vulture spends much of its life soaring in search of carrion.

The turkey vulture holds a special role in American history because Wilbur Wright studied the bird's dihedral flight to help learn how to achieve stability when designing the Wright brothers' first successful airplane.

A soaring turkey vulture inclines its wings slightly upward. This posture results in a distinctive shallow V-shape of the wings held above the back, creating what's known as a dihedral angle. This posture provides a significant flight benefit. Whenever a wind gust causes one of the vulture's wings to tip downward, that wing generates more lift. That is because more lift is generated along the greater length of the horizontal wing than by the wing that is tilted upward. This causes a "self-righting" effect that pushes the lower wing back up to the dihedral position.

In Africa, the bateleur is an eagle that flies about two hundred miles per day in search of small creatures and carrion. It has a strong, direct gliding flight characterized by a dihedral wing posture. The name bateleur derives from a French word meaning "acrobat." Near its nest, the bird puts on a display, tumbling over and over in the air. While in flight, the constant tipping action of the dihedral wings are reminiscent of the tipping arms of a tightrope walker. Some other birds, such as terns and pigeons, also use the dihedral position while momentarily gliding, but they do not glide for extended periods.

3. Long, Angular, Pointed, High-Speed Wings

Falcons, swallows, swifts, boobies, gannets, frigatebirds, kites, gulls, and plovers are adapted for life in the fast lane. They have long, medium-width wings with a small point at the wingtips. The points minimize drag created by air turbulence at the wingtips. The wings of these birds are characterized by low wing loading and a high aspect ratio. Many of these birds carry out extremely long migrations or spend extended times aloft in search of prey. Their flight is characterized by a combination of flapping, gliding, and soaring to minimize the energy necessary for flight.

Long, angular, pointed wings allow great agility. Many of these birds capture live prey on the wing either in the air or at the water's surface. Most of these birds also have an angular profile at the wrist joint.

The slightly arched surface of a feather is called the camber. The primary feathers of these birds are relatively flat (low camber) compared to the strongly concave profile (high camber) of turkey and pheasant primaries. They do not need the extreme takeoff speed provided by high camber wings because they are adapted primarily for soaring.

Flight for some of these birds can include dramatic aerial maneuvers. Swallows and swifts capture insects in their open mouths while on the wing. Peregrine falcons have a wing design that allows them to reach speeds up to two hundred miles per hour as they dive toward birds in flight. Swallows and swifts are extremely adept fliers and spend most of their lives flying—up to sixteen hours per day. Some swifts even sleep while flying. Ornithologists estimated that a nine-year-old banded chimney swift, having likely migrated nine times from North America to Brazil and back, had flown 1,340,000 miles during its life.

The chimney swift has long, pointed wings that are well-adapted for fast flight and excellent maneuvering ability as it pursues flying insects. Its wings are also effective for long-distance flight, allowing the swift to migrate from North America to Brazil.

The long, slender, pointed wings of the barn swallow enable it to fly swiftly and agilely and capture insects on the wing. Its wings also allow the swallow to migrate thousands of miles every fall from North America to wintering areas in Central America.

The negative dihedral contour of this great frigatebird allows it to carry out agile dives, turns, and other aerial maneuvers necessary to pursue other Galápagos seabirds. The harassed seabirds then drop their food, which the frigatebird steals in midair.

Negative Dihedral Wings

One of the most impressive avian flight profiles is that of the frigatebird. It flies with its wings in a negative dihedral angle, meaning it inclines the proximal portion of the wing with the secondaries slightly upward and the extremely long primaries slightly downward. Aviators consider this design aerodynamically unstable. However, among frigatebirds, it is an adaptation for extreme maneuverability when diving, swooping, and chasing other birds.

The frigatebird is the stunt pilot of the bird world. The negative dihedral angle of the wings combined with the long forked tail gives the bird exceptional maneuverability in the sky. Also known as the pirate of the air, it uses its skills to harass boobies and tropicbirds until they drop or disgorge fish that they have previously captured. Then the frigatebird acrobatically catches the falling fish before it hits the water.

The frigatebird has an interesting limitation to its foraging abilities. It is unable to land on water to feed because its feathers would become waterlogged. It could not take off from the surface of the water. It does not feed on land. All of its food must be taken on the wing by capturing it at the water's surface or by stealing it from other birds in flight.

4. Short, Broad, Rounded Wings of Fowl

Wild fowl such as grouse, pheasants, francolins, ptarmigan, guinea fowl, and turkeys have short, broad, rounded wings designed for quick takeoffs and strong, rapid, relatively short flights. The wings are characterized by high wing loading and low aspect ratio. Fowl live and nest on the ground, where they feed primarily on seeds, fruits, and insects. Their legs are strong and well-developed because they move about primarily by walking or running. These birds serve as prey for foxes, coyotes, jackals, hawks, and owls. Their short, broad

Short, broad wings with rounded wingtips are characteristic of wild fowl such as pheasants, quail, grouse, and francolins. A pair of fighting male crested francolins of Kenya demonstrate their ability to explosively leap from the ground and become airborne.

wings allow them to flush in a surprising and noisy manner that can startle a predator and cause it to miss its quarry. If a predator disrupts a covey of fowl, the loud noise of a single flushing bird alerts the other birds to the presence of danger.

The wings of fowl are strongly cupped like an upside-down saucer. The high camber traps air under the wings when flushing, so the birds are more efficient in making a quick getaway. These birds, however, are adapted best to short flights of less than a mile and most are not migratory.

This wild turkey wing is significantly curved. In aerodynamic terms, it has a high amount of camber. High camber helps the bird flush explosively from the ground.

5. Short, Broad, Rounded Wings of Songbirds and Parrots

Most songbirds, doves, flycatchers, parrots, woodpeckers, and other smaller birds have wing profiles that allow them to flush and fly away quickly. These birds typically inhabit forests or grasslands and nest in trees or on the ground, and their wing type helps them escape birds of prey. Unlike fowl that are adapted for explosive flushes and short flights to nearby cover and feeding areas, songbirds generally have great flight endurance. Their wings typically have a low aspect ratio and low to medium wing loading.

Many songbirds migrate thousands of miles each year. The bobolink and the dickcissel—songbirds named for their calls—migrate from North American grasslands to wintering grounds in South America. Many North American warblers migrate from forests of the United States and Canada to tropical forests of Central America and islands of the Caribbean. A few continue on to the rainforests of South America.

Birds of the parrot family do not migrate, but they cover many miles every day in their tropical forest habitats searching for ripe fruits and seeds. They also may travel great distances to reach clay licks along rivers. Eating the fine-grained sediment helps neutralize potentially toxic chemicals in some of the seeds they consume. Parrot wings are relatively short and wide, giving them a low aspect ratio. This design enables them to flush quickly if pursued by a raptor and provides for rapid flight through the forest canopy in their constant search for food.

Slotted wingtips are conspicuous on large birds such as hawks, eagles, cranes, and vultures. However, close examination of songbird wings, like those of the indigo bunting and common redpoll, reveal that they also have slotted wingtips. The slots are either V-shaped, U-shaped, or square notches between the tips of the primary feathers. This characteristic helps reduce turbulence at the wingtip and make flight more energy-efficient.

The short, rounded wings of the red-winged blackbird bear superficial resemblance to those of a grouse or pheasant, but they are used in a quite different manner. The blackbird's wings are used for agile flight among trees and shrubs as well as in open habitats. They are also adapted for long-distance migrations, whereas wild fowl are not migratory.

Songbird wings such as this indigo bunting wing have slots between the distal primary feathers. These V-shaped slots serve as small airfoils that help generate a smooth airflow at the wingtip. Without the slots, the wing would experience more air turbulence.

The wings of the green hermit hummingbird exemplify the straight, narrow, oarlike structure of a typical hummingbird's wing. The wings may beat fifty or sixty times per second in order for the bird to hover.

6. Long, Slender, Straight, Oarlike Wings

The hummingbird has long, slender, straight wings that "row" the air like the oars on a boat. It does not glide, soar, or flap its wings, nor does it generate lift with the secondary feathers or thrust with the primaries. Instead, the unique hummingbird beats its wings as many as two hundred times per second, allowing it to pollinate flowers while hovering.

Unlike spinning helicopter blades, hummingbird wings provide lift on both the forward stroke and the backward stroke. It works like this: From a starting position in which the wings are pointed backward, the hummingbird propels the wings forward, generating both lift and thrust. At the end of each forward stroke, the humerus rotates backward in its socket and the wings flip upside down. During the backstroke, the wing's front edge is slightly higher than the rear edge and therefore generates lift.

By adjusting the posture of the wings, the hummingbird is able to fly up, down, forward, backward, and even upside down. It is the only bird able to fly backward.

BLACK WINGTIPS

In 1923, ornithologist Charles K. Averill recognized that primary feathers with black tips resisted wear better than white feathers and published an article about this phenomenon. Indeed, a high frequency of birds have black wingtips. Examples include American white pelicans, snow geese, gannets, boobies, storks, ibises, and gulls. This is not an accident. The black coloration is caused by melanin pigments that contribute to feather hues ranging from dull yellow and reddish-brown to brown and black. The tips of a bird's primary feathers face the greatest stress from flight because of the high air speeds to which they are exposed. Since primary feathers must typically last for a year before molting, birds would be at a great disadvantage if those feathers wore out prematurely.

The primary feathers of the ring-billed gull offer an interesting study in feather wear and the functions of melanin. The image shown here (*below, top*) portrays a new, unworn fall feather. It still has a white tip and black edging along the trailing edge of the feather. The other feather tip (*below, bottom*) shows the wear after about a year of use. The white tip has broken off and part of the black edging has broken off where it joined the white coloration, creating a notch in the feather. This raises an interesting question: Why would gulls have such a "flaw" that causes part of the tips to break off prematurely?

A possible answer is that broken feather tips create notches that function like slotted feather tips and reduce turbulence.

In contrast, large white aquatic birds such as swans have white wingtips, making them among the only birds to have this physical characteristic. The trumpeter swan, tundra swan, and jabiru all have white wingtips. Apparently, the feathers on these birds are so large and durable that the feather tips do not wear out as readily as they do on smaller birds.

Opposite: The Australasian gannet *(top)* and sacred ibis *(bottom)* are from widely separated places around the world— New Zealand and Africa. However, they both have black wingtips that give strength and durability to their primary wing feathers.

Right: The primary feathers of the ring-billed gull demonstrate the weakness of white feather tips. The top feather is newly grown and does not show extensive wear; the bottom feather was shed after about a year of use. Notice that after the black sections have broken off, the feathers have new slots to reduce turbulence.

Most birds that fly in formation are waterbirds: ducks, swans, geese, and pelicans. These are tundra swans.

MIGRATION FORMATIONS

Why do birds often migrate in lines or V-shaped formations? Ornithologists are not in total agreement about the answer. The most accepted theory is that the trailing bird benefits from the updraft of air created by the wingtip vortex of the forward bird. The trailing bird flies so that one of its wings is within the pocket of updrafting air. Even more intriguing is that some birds, such as pelicans, beat their wings in unison while in formation to take maximum advantage of the updrafts created by the wingtip vortex.

Ducks and geese may be better known for flying in V-shaped formations, but these lesser flamingoes of Kenya also take advantage of the lift provided by the vortex at their wingtips to save energy during flight.

An alternative theory is that the air disturbance created by the leading bird actually creates an area of reduced pressure that helps propel the trailing bird forward, more so than just a vortex pocket behind the wingtip.

Migrating birds also fly in the standard V-shaped formation because the trailing bird can see the flock and where it is going when it is just outside the forward bird. If the trailing bird were to fly to the inside of the V, its only view would be the tail end of the bird two positions ahead in the flock. Military planes fly in a V-formation for the same reason.

The lead bird in the V-shaped flock expends more energy than the other birds in the formation. As it eventually tires from being in the lead, it will drop back and let another bird take over the lead position.

THE TALE OF THE *Tail*

Most people do not think of tails when they think of bird flight. However, tails are important for carrying out sophisticated aerial maneuvers. The tail is a bird's rudder, much like the tail of an airplane. When spread wide in a horizontal plane, the tail keeps the bird level. When the tail is raised, the bird ascends. Lowering the tail helps the bird dive. When the tail is sharply and abruptly lowered, it brakes forward movement and aids in landing. Finally, spreading the tail and sharply twisting it to the right or left helps the bird bank sharply to the right or left, respectively.

A bird's tail makes it possible to carry out acrobatic and skilled maneuvers while pursuing prey or avoiding predators. *RTimages, Shutterstock*

As a blue and yellow macaw flies in to a clay lick along the Tambopata River in Peru, its tail plays an important role in helping the bird brake while landing.

Climb:
Elevator up

Dive:
Elevator down

Tail feathers are seldom soft and fluffy, except on a few flightless or nearly flightless birds such as ostriches and members of an African rail family known as flufftails. Most birds have twelve tail feathers; a few, including hummingbirds and swifts, have ten tail feathers. Prairie chickens and ring-necked pheasants have eighteen tail feathers, and pelicans have twenty to twenty-four tail feathers. Pheasants and macaws have extremely long, ornate tails that are used for courting and also for braking and maneuvering in flight. The blue and yellow macaw *(above)* is using its tail to brake as it approaches a clay lick in eastern Peru.

To provide precise flight movements, tail feathers need to be stiff in order to assist with taking off, turning, and landing. The length of a bird's tail correlates to adaptations to its habitat. A bird that lives in woodlands, like a grouse, needs a high degree of maneuverability to fly around trees and bushes. A medium length to long tail aids in this. Woodland raptors also need medium length to long tails to help them pursue prey such as grouse or rabbits zigzagging among trees and shrubs.

Birds that inhabit open grasslands, wetlands, or oceans have shorter tails, including ptarmigans, prairie grouse, francolins, gannets, and albatrosses. They do not need a high degree of maneuverability in their open habitats. Several types of tails are described here as examples of the diversity in form and function.

Airplane Rudders

This stylized diagram of two airplanes shows how the rudders on the tail are used to help the plane dive or climb. When the rudders are tilted up, the airplane climbs. When the rudders are tilted down, the airplane dives. The same principles apply to how a bird's tail controls these movements and aids in turning right or left.

HUMMINGBIRDS

The hummingbird tail provides an excellent example of maneuverability. The bird manipulates its position with high precision as it approaches flower blossoms, goes through complex high-speed courtship rituals, and fights for feeding territory among flower blossoms. By lowering the tail and spreading the tail feathers, the hummingbird can stop abruptly as it hovers. It can also snap the tail downward and outward to stop and then change directions.

A Brazilian ruby demonstrates the amazing maneuverability provided by its tail. The hummingbird has flipped its tail downward and forward and thrust its body backward so that it can pivot and reverse its flight direction.

The elegant flight and conspicuous forked tail of the swallow-tailed kite make it one of the most distinctive raptors in the Americas. When the tail is spread, it increases drag and allows the kite to slow down, bank, or turn with great agility as it picks cicadas, small lizards, and snakes from the foliage of tropical treetops.

A forked tail can be spread to slow a bird's flight or provide abrupt turning ability. This common tern can also fold its tail straight back to minimize drag during its long migrations between North and South America. *Carrol Henderson, Minnesota Department of Natural Resources*

FORK-TAILED WONDERS

Among the most graceful and elegant of all birds are those that possess forked tails. Frigatebirds, swallow-tailed kites, common terns, Forster's terns, and barn swallows are all examples of birds with forked tails. Other species include scissor-tailed and fork-tailed flycatchers and swallow-tailed humming-birds. Fork-tailed birds typically pursue prey in the air or at the water's surface using maneuvers like twisting, turning, banking, and diving. When its forked tail is spread wide, a bird exposes a broad horizontal surface that helps maintain a level posture. The outer feathers of the outspread tail also can assume the role of an extra set of wings—like the second set of wings on a biplane. These offer extra profile drag, stability, and maneuverability while slowing flight speed. These features aid the bird in making sharp turns and holding its position while watching for prey below.

Common terns and barn swallows are both examples of fork-tailed birds with excellent maneuverability. The tail provides great maneuverability as the bird swoops and dives for prey either in the air (in the case of the barn swallow) or at the surface of the water (in the case of the common tern). The tail helps a tern maintain balance when it "wind hovers" while watching for minnows. Wind hovering is a flight behavior in which the bird faces into the wind and uses its wings to hover in place as it searches for food below. You can find more information on wind hovering in Chapter 7.

The red-billed tropicbird of the Galápagos is closely related to the pelican. Its streamertail is used during courtship displays and, in flight, contributes to longitudinal stability.

STREAMERTAILS

Another tail design is the streamertail. This extravagant tail comprises at least two long central feathers that extend two or three body lengths behind the bird. Red-billed and white-tailed tropicbirds and violet-tailed and long-tailed sylph hummingbirds have streamertails. It often serves as part of courtship displays. The long narrow tail also functions as a longitudinal stabilizer much like the tail on a child's kite. The tail of the tropicbird helps the bird maintain stability as it captures minnows at the ocean's surface and as it navigates updrafts when approaching its cliffside nest. The stringlike tail of the sylph hummingbird helps it maintain position as it obtains nectar from the tiny openings of rainforest flowers.

The bateleur is an eagle with an extremely short tail. Because it feeds mainly on carrion in the open grasslands of Africa, it does not require a long tail that would provide agility for pinpoint landings or capturing live prey.

STUBTAILS

Some birds have short, inconspicuous tails, including albatrosses, quail, ducks, prairie chickens, ibises, and some vultures. There is a logical explanation. Presumably there is an adaptive "cost" to longer tails for these species, and natural selection has eliminated individuals with longer tails. Most live in open environments where extreme maneuverability is not necessary. The bateleur eagle has an extremely short tail and a strong gliding flight that allows it to fly over the African savanna looking for carrion or small birds, mammals, and reptiles. Agility is not a requirement for obtaining carrion as food. The green ibis of Central America has a short tail even though it lives in forested swamps. However, the wings have a low aspect ratio that provides good maneuverability and braking and turning potential in the swamps and waterways.

The impressive wedge-tailed eagle is the largest eagle in Australia, known for its stunning aerial displays during the mating season. The male dives toward the female and swoops up after passing near her. He may "loop the loop" in his attempt to impress the female. The eagle's tail plays a major role in carrying out these maneuvers. *Lee Torrens, Shutterstock*

WEDGE TAILS

One additional tail design is the wedge or paddle-shaped tail. Examples of birds with wedge tails are wedge-tailed eagles, Egyptian vultures, Brahminy kites, sea eagles, and Neotropic cormorants. The Neotropic cormorant, an aquatic diving bird that inhabits tropical wetlands, has a wedge tail with a narrow muscular base. The wedge tail provides considerable maneuverability for landing and taking off among the tropical trees and shrubs where the cormorant roosts and nests. The tail can twist sideways to facilitate abrupt turns or spread out to maintain horizontal flight stability. Together with the cormorant's webbed feet, the paddle-shaped tail design is an important adaptation for swimming underwater and allowing the bird to twist and turn as it pursues fish.

The wedge tail of the Neotropic cormorant serves as a rudder in flight and while pursuing fish underwater.

PART III

On the Wing

Above: A red-tailed hawk takes off by pitching forward from a tree branch and gaining enough airspeed while dropping to become airborne.

Left: As Ciorella cockatoos of Australia take flight, they demonstrate their athletic ability to leap into the sky at a moment's notice. *M. Willis, Shutterstock*

6

TAKING *Flight*

The expression "taking flight" creates vivid images of birds jumping, flapping, running, and splashing their way into amazing and graceful aerial performances. When a bird takes flight, it must achieve enough airspeed for the lift to equal the force of gravity. Different birds accomplish this in different ways. However, taking flight always involves the basic principles of aerodynamics relating to thrust, drag, lift, and gravity. For humans, who can only dream of such maneuvers, flight is a phenomenon still approaching a miracle. Within my collection of bird photos, I have identified five different ways in which birds gain flight: 1) dropping forward and down, 2) flushing upward and/or forward, 3) jumping upward, 4) running into the wind, and 5) lifting off upward and backward.

As you review the various ways in which birds take flight, think about how each takeoff technique is specifically adapted to the lifestyle of the bird in question. For example, a trumpeter swan requires a long run across an open surface of land or water and would not survive long in a brushy, forested habitat.

Avian flight involves a range of aeronautical skills: taking off, full flight, descending, and landing. *Mark William Penny, Shutterstock*

The bank swallow takes flight by dropping forward and down from its nesting cavity. Gravity helps the bird reach the critical speed at which it spreads its wings, generates lift, and becomes airborne.

1. Dropping Forward and Down

Swallows, martins, and tropicbirds have small, weak feet that make them incapable of jumping into the air or running across the ground or water to take off. The bank swallow takes flight from a burrow in a cut bank by pitching forward with wings outspread so that as it falls forward it immediately reaches enough speed to gain flight.

The tropicbird nests on cliffs by the ocean where it also pitches forward into the sky with wings outspread and immediately gains enough momentum to fly. Hawks may also pitch forward and down from elevated perches to become airborne.

2. Flushing Upward/Forward

Many songbirds, fowl, and doves take flight with a powerful downstroke of the wings. Their legs are not particularly strong, so they depend primarily on their wings to lift them into the air. These birds may flush from a perch in a tree or from the ground. One unique feature of this technique is that when the wings are uplifted and then thrust sharply downward, the area above the bird's back becomes a partial vacuum, and the bird's body is pulled upward into the vacuum. This helps the wingtips avoid slapping the ground or perch as the bird flushes.

Songbirds like the gray-backed camaroptera of Kenya flush from their perches with a strong downstroke of the wings.

Terns and waterfowl known as puddle ducks are also examples of birds that flush upward from the water with a powerful downward thrust of the wings. Their feet do not play a significant role in flushing. Puddle ducks include mallards, northern pintails, blue-winged teal, northern shovelers, wigeons, and gadwalls. They generally have a low aspect ratio and low to medium wing loading. They can flush with considerable speed and agility. This is important because most of these birds serve as prey for raptors and terrestrial predators.

The northern shoveler of North America flushes upward from the water with a powerful downstroke of the wings.

3. JUMPING UPWARD

Herons, egrets, ibises, and hawks have more powerful legs than the birds previously mentioned. Their legs play an important role in taking flight. They leap upward as they take flight with a powerful downward thrust of their wings.

When red-tailed hawks *(left)* and great blue herons *(above)* take flight, they usually leap upward with their legs and become airborne using powerful wingbeats.

When the common loon takes off from the water, it must overcome high wing loading by running into the wind while flapping until it achieves enough speed and lift to become airborne.

4. RUNNING INTO THE WIND

Many birds face into the wind as they flush because it facilitates becoming airborne. The stronger the wind, the easier it is to achieve the lift necessary to take flight. This is where the concept of high wing loading becomes important. The common loon has high wing loading. Its body weight is fairly heavy considering the surface area of the wings. It must generate both lift and thrust with the aid of strong winds or a high takeoff speed. A loon faces into the wind and begins flapping its wings and paddling with its feet, then running across the surface in what appears to be a frantic sprint into the sky. Water droplets kicked into the air appear as a crystal spray as the loon gains

speed and lifts off the water. Depending on the wind speed, it may take a loon fifteen yards to more than a hundred yards to become airborne. Other waterbirds also take off by running across the water to become airborne, including grebes, mergansers, and swans.

Different species of waterbird are adapted for different lifestyles. Puddle ducks spend time on both water and land and flush upward to take flight, whereas diving ducks live an almost exclusively aquatic existence and must run across the surface of the water into the wind when taking off. Diving ducks include canvasbacks, redheads, scaup, scoters, eiders, buffleheads, ring-necked ducks, and ruddy

The southern royal albatross has extremely high wing loading and must take off into strong winds. The strongly arched wings help generate lift as the albatross runs along the water's surface into the wind.

A ruby-throated hummingbird becomes airborne by hovering on its perch and then lifting upward and backward.

ducks. Their relatively small wings have high wing loading, and their feet are webbed and located farther back than on puddle ducks. Their physical characteristics lend them to an aquatic existence but it also means they do not walk well on land. They nest right at the shore or over water and feed by diving underwater and nabbing fish and other aquatic organisms.

A wind-based takeoff would be a liability for many land birds because most upland habitats have trees or shrubs that would obstruct a "runway" for taking off. However, birds of open grasslands and marshes, such as cranes, run into the wind to take off.

The albatross presents an extreme example of needing to take off into the wind because of its high aspect ratio and high wing loading. First, it faces into the wind and lifts its wings into an impressive arching posture that begins to create lift. Then it begins a ponderous run into the wind. If the wind is strong enough to generate the necessary lift, the albatross becomes airborne. If the sea is quiet, the albatross may not be able to take flight.

5. LIFTING UPWARD AND BACKWARD

When a hummingbird lifts off from a twig perch, it hovers so that the lift carries it upward and backward while it releases its grip. Hummingbirds have feet, but they never walk, run, or hop. Their feet function only for perching. For example, if a hummingbird wants to move two inches to the left on a twig, it lifts off, hovers to the left, and alights again.

7

Types OF FLIGHT

The aspiration to fly like a bird has long been a part of human culture. Greek mythology includes the story of Icarus and Daedalus whose attempt at flight failed when Icarus, wearing large bird feathers attached to his arms with wax, flew too close to the sun. When the sun's heat melted the wax, Icarus fell to his death.

Understanding avian flight is an exercise in bird body language. This double-crested cormorant does not look like it should be able to fly, but its upright posture is necessary for increasing drag and decelerating just prior to landing. *Larsek, Shutterstock*

Leonardo da Vinci was famous not only for painting the *Mona Lisa* and other classic works but also for his infatuation with avian flight. He studied birds and attempted to design a device, which he called an ornithopter, that would allow humans to fly. Unfortunately, da Vinci and other early students of flight thought of bird flight as a pattern of flapping wings up and down. However, simply flapping up and down does not generate aerodynamic benefits resulting in flight.

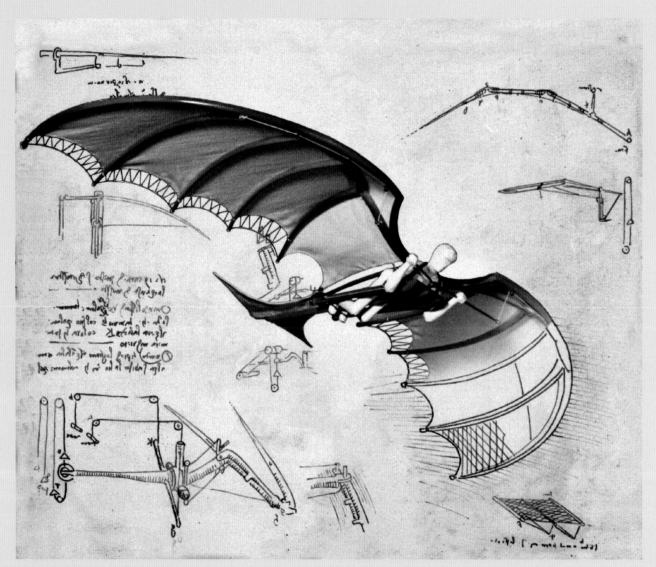

Leonardo da Vinci conceived the ornithopter in the fifteenth century. It was intended as a means for humans to achieve flight by flapping their arms within the device.

From Fledgling to Flapping

In 1966, the staff at the Lac qui Parle Wildlife Management Area near Watson, Minnesota, discovered the nest of an American white pelican. It was the first nesting record for American white pelicans in Minnesota in almost ninety years! Each year thereafter, the number of nests increased. When I arrived to work at Lac qui Parle in 1974, hundreds of pelicans were nesting. During a July visit to their nesting island in Marsh Lake, I viewed the prehistoric-looking, flightless chicks for the first time. Looking upon such helpless creatures, I saw no hint of any potential for flight. No aerodynamic design presents itself in those naked little creatures. Within only three months, however, I saw them transform into one of the most graceful and magnificent of all flying birds. Their soaring flight seemed so effortless. As I watched their majestic figures in the sky, it gave me a new appreciation for avian flight.

The American white pelican undergoes a remarkable transformation from a fat, naked hatchling *(above)* into one of the most graceful flying birds of North America *(below)*. By taking advantage of its high aspect ratio and low wing loading, the pelican is able to soar on thermal updrafts for extended periods of flight. *Carrol Henderson, Minnesota Department of Natural Resources*

Most people think of flapping wings when they think of bird flight, but birds actually use about six different types of flight: flapping, gliding, soaring, hovering, wind hovering, and pattering. Each species may use more than one of these techniques.

These compound photographs, from left to right, show four consecutive wing positions of a great egret in flapping flight.

FLAPPING FLIGHT

Flapping flight is one of the most widely used techniques among birds, ranging from tiny house wrens to enormous trumpeter swans and great bustards. The apparently simple up-and-down movements of the wings belie the complexity of this form of flight. Some fifty different muscles control the flapping of a bird's wing, so the movements are extremely sophisticated.

As a great egret flies, the primaries are driven down and forward in a powerful downstroke. This

creates a propeller-like forward thrust. At the same time, the secondaries are held in a nearly horizontal position and do not pass through as large an arc as the primaries. The secondaries generate lift in their horizontal position while the primaries generate thrust with the fuller downstroke.

Since each wing passes through the downstroke simultaneously, the egret essentially has a pair of reciprocating propellers that each pass through an arc of about 50 to 60 degrees with each wingbeat.

At the completion of the downstroke, the humerus is drawn in toward the body and then upward in preparation for the next downstroke. Individual primary feathers then separate and flip backward while in an upside-down position. This phenomenon is described as the venetian blind effect in chapter 2. The leading edges of the primary feathers are slightly upturned to create an angle of attack that generates some lift during the backstroke. All of these subtle changes take place during just one flap of a bird's wings.

For smaller birds, like warblers and tanagers, the wingbeat is more rapid. Also, compared to a goose or swan, the entire wing is more involved with each flap and goes through a greater arc.

Swallows, martins, nighthawks, and swifts are impressive hunters that capture their insect prey on the wing while using flapping flight. They spend a considerable amount of their lives in the air as they pursue insects and catch them in their beaks. Martins prefer to pursue larger prey such as dragonflies and damselflies over water or near wetlands. Swallows, swifts, and martins capture day-flying insects.

Most birds flap their wings synchronously to achieve level flight. The swift is one of the only birds that sometimes flaps one wing at a different rate than the other. Flapping the right wing more rapidly than the left wing facilitates turning to the left, and visa versa. Because swifts have extremely short tails, flapping the wings at independent rates may assist in turning or performing other aerial maneuvers that would otherwise be difficult with a short tail.

The nighthawk takes flight at dusk to feed on moths and other nocturnal insects. Capturing insects in its open mouth takes great agility. To facilitate this feeding technique, the nighthawk's wings are long and angular. This design helps the nighthawk fly at high speeds and perform acrobatic turns, twists, and dives. The medium-length tail aids in making sharp turns. A single bird may take hundreds of insects every night to satisfy its dietary needs. In the fall, the nighthawk must leave northern regions when its insect prey becomes unavailable. The bird undergoes a long migration to Central and South America, so its wing designs must serve it well for long flights as well as for agility in catching insects.

Opposite: The prothonotary warbler utilizes flapping flight, but the wingbeat is rapid and virtually impossible to discern with the human eye.

The mangrove swallow utilizes rapid flapping flight and possesses a great agility to capture insects on the wing. Swallows may be small birds, but they are capable of long migrations.

The common nighthawk has long, pointed angular wings adapted for swift and agile flight in order to catch insects on the wing.

Gulls and terns, like this sandwich tern, are skilled at gliding because their wings provide considerable lift when held in an outspread position.

When this Peruvian pelican flies just above the water, it is riding on a cushion of air that is momentarily compressed between its body and the water's surface.

GLIDING

Gliding is defined as a continuing descent from a higher to a lower altitude with no wing flapping and no use of thermals or updrafts to extend the distance traveled. The gliding range of a bird can be extended after a long descent by flapping to a higher elevation and then gliding downward again. Most birds can glide, if only momentarily, except hummingbirds.

Glide ratio measures how far a bird can glide forward for each foot of descent. The more efficient the aerodynamic structure of the wings, the farther a bird can glide forward. The albatross is the most impressive gliding bird in the world. It has a wingspread of nearly twelve feet and weighs up to twenty-two pounds. Because of its great wingspan, the wandering albatross has a long span of secondary feathers providing lift and the most impressive glide

ratio of any bird. It drops only a foot in elevation for each twenty-three feet that it glides forward. In contrast, a common tern drops a foot for each twelve feet that it glides forward, and a house sparrow drops a foot for each four feet of gliding flight.

Air Cushion Gliding

The brown pelican is a master of gliding on a cushion of air. When a pelican flies low over the ocean—just inches above the water—the air becomes compressed between its body and the surface of the water. The compressed air generates more lift than would be possible at an elevation farther above the water. The extra lift facilitates gliding long distances over water with a minimum of wingbeats and is called "air cushion gliding." Shearwaters also utilize this form of flight.

The Andean condor is master of its domain. It soars for hours on updrafts amid the peaks and canyons of the Andes Mountains in South America.

SOARING

Soaring is similar to gliding, but instead of descending, the bird sustains its elevation or rises to higher levels by taking advantage of updrafts. These updrafts may be created by ascending thermals of warm air or by sloping terrain or cliffs that drive air currents upward.

The vulture, one of the most well-known soaring birds, may fly for hours as it searches for carrion. From its high vantage point above the land, it looks for dead and dying animals that will provide a meal. A soaring vulture watches the behavior of others of its kind. Once vultures begin to descend, their flight alerts other vultures of available carrion.

Condors are the largest of all vultures. Two species, the California condor and the Andean condor, inhabit North and South America. The Andean condor is larger and has a wingspread up to 10.5 feet. It can be found from sea level up to elevations over 20,000 feet in the Andes Mountains. Watching the graceful and majestic flight of condors is awe-inspiring.

Kites, hawks, and eagles do not typically soar to the extent of vultures or frigatebirds, but they do have wings with a high aspect ratio. Strong and steady winds allow them to take advantage of the considerable lift generated by their wings. With wings fully extended, they appear to float through the sky.

Thermal Updraft Soaring

A thermal is a large upwelling current of warm air rising sometimes thousands of feet above the ground. It is a "bubble" of warm air rising in a sea of cooler air. Some of these bubbles are large, and others are smaller. Thermals are created over areas of bare ground, grasslands, rocky outcrops, or over urbanized areas characterized by concrete and asphalt. When these areas become warmer during the day, a column of warm air rises, like a bubble, and then disperses at higher elevations in a mushroom-shaped pattern as the warm air at the top cools and then descends.

Thermal Updraft

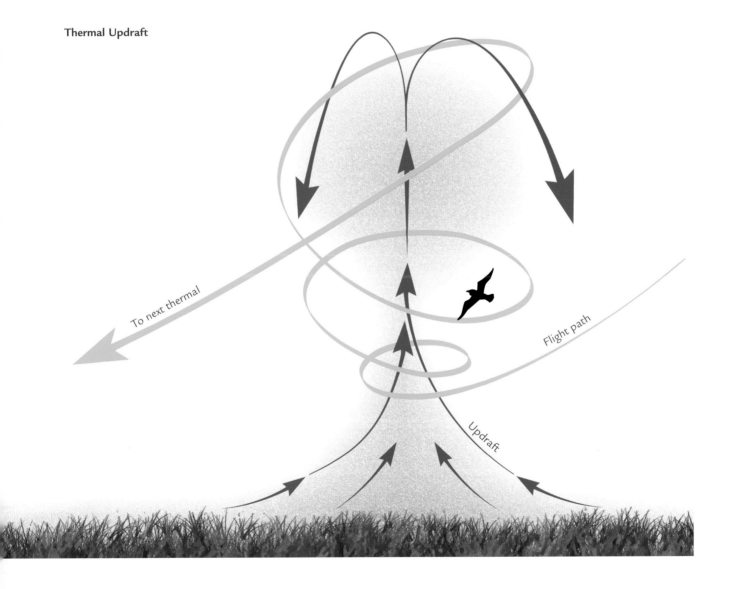

Thermals help soaring birds migrate effortlessly for hundreds of miles. They take advantage of multiple updrafts along their migratory corridor. For example, black storks migrate every March from their wintering grounds in Kenya to their nesting areas in Europe. As air warms over rocky areas in the Kenyan plains of the Masai Mara, the rising air columns are used by black storks, who circle effortlessly, higher and higher within the columns. When they reach the upper limit of an updraft, they leave and glide many miles to the north in search of another updraft. In this manner, they use less energy in their migration back to Europe.

Black storks soar on thermal updrafts above the Masai Mara in Kenya. It is March, and they are migrating north to return to their breeding grounds in Europe.

Half a world away and also in the month of March, Swainson's hawks begin their annual spring migration from the pampas of northern Argentina to the Great Plains of the northern United States and Canada. As they pass over the shrubby grasslands of eastern Bolivia, they soar on thermals near Santa Cruz at the foot of the Andes Mountains. A concentration of soaring hawks circling higher and higher on a thermal is called a kettle. The soaring hawks can be so numerous and so high that they look like a cloud of mosquitoes. From one point, birdwatchers can observe several kettles above the grasslands, each

Left: The Swainson's hawk carries out a long annual migration route between the Great Plains of North America and the pampas grasslands of northern Argentina.

Below: An Australasian gannet of New Zealand uses rising air currents along a cliff face for additional lift while soaring.

Slope Updrafts

Flight path

Prevailing winds

working its way north from one thermal to another. Eventually, the hawks migrate through Venezuela, Panama, Mexico, and on to the Great Plains. When our Henderson Birding group photographed this spectacle of migrating Swainson's hawks near Santa Cruz, Bolivia, in 1996, we learned that this site was the first migratory hawk corridor that had been discovered in all of South America.

The pelican counts among the most graceful of all flying birds. It has a high aspect ratio and low wing loading. With wings fully outspread, it appears to float through the sky. The American white pelican uses thermals while migrating between its northern nesting grounds and wintering grounds along the Gulf of Mexico.

Updrafts are not unique to wilderness areas. Heat rising from cities generates urban thermals. In downtown St. Paul, Minnesota, city-dwellers can see kettles of American white pelicans circling over the city in the spring as they migrate to nesting areas in northwestern Minnesota, North Dakota, and Canada.

Slope Updraft Soaring

A slope updraft is a current of air rising along the slope of a hill, mountain, or cliff face. It presents another opportunity for soaring raptors to ride the rising air in an effortless manner. Hawk Mountain in Kempton, Pennsylvania, and Hawk Ridge in Duluth, Minnesota, are famous migratory corridors where raptors take advantage of updrafts along the hillside topography as they migrate south in the fall.

Prevailing sea winds blowing off the ocean and coming in to adjacent hills create slope updrafts used by terns, tropicbirds, and soaring raptors including vultures. At Turtle Mountain near Tortuguero, Costa Rica, vultures end the day by soaring on the slope updrafts up the small mountain to their nocturnal resting areas.

Slope updrafts in Costa Rica have also been observed to be used by kettles of white-collared swifts, which fly in circular formations late in the day prior to roosting. The swifts apparently benefit from the slope updraft of the hill while they move to their night roost site.

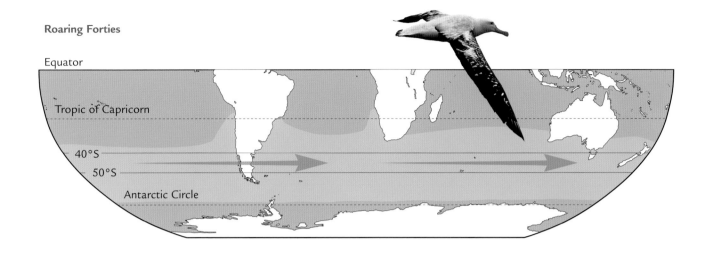

Roaring Forties

Equator

Tropic of Capricorn

40°S
50°S

Antarctic Circle

The wandering albatross utilizes dynamic soaring as it takes advantage of high winds in the Roaring Forties.

Dynamic Soaring

An oceanic region known for its consistently high winds and called the Roaring Forties is located in the Southern Hemisphere between 40 and 50 degrees latitude. In addition to the winds, the cold water in the region hosts an abundance of squid, fish, and marine invertebrates. These are important foods for seabirds. In the Roaring Forties, oceanic winds blow slower near the water's surface and are progressively stronger with increasing elevation. Wind speeds reach their maximum velocity about one hundred feet above the water.

The albatross has an elliptical, rollercoaster manner of flying referred to as dynamic soaring. Starting at an elevation of about one hundred feet,

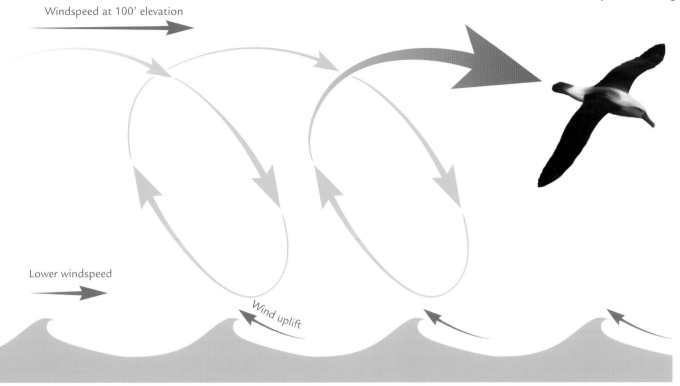

Windspeed at 100' elevation

Lower windspeed

Wind uplift

an albatross will glide downwind as if on a roller-coaster, gaining velocity from the tailwind as it approaches the ocean's surface. On nearing the water's surface, the bird lowers one wing, banks, and then soars back up at an angle to the wind. The albatross uses its increased velocity to soar upward, angling into the wind, back to an elevation of about one hundred feet, where it turns and repeats its spiraling glide routine. Depending on which way the albatross turns at the bottom of its flight, it flies in either a circling or an undulating pattern as it works its way east with the wind. Dynamic soaring requires virtually no wing flapping. The albatross is able to "lock" its wing bones in place as it soars. Albatrosses have been known to traverse 19,000 miles

around the world in the windy Roaring Forties in only forty-six days. Wandering albatrosses are the greatest travelers among all birds. It is estimated that a fifty-year-old albatross has traveled 3.7 million miles during its life.

One of the best places in the world to watch albatrosses is at Kaikoura in New Zealand. Kaikoura is on the east coast of the South Island. Since the continental shelf lies only a half mile from the city, the eternally soaring albatrosses pass near land. In several hours of boating, a birder might see wandering, Buller's, northern royal, southern royal, black-browed, and white-capped albatrosses. Another excellent albatross-watching site is offshore from Stewart Island, south of New Zealand's South Island.

Hovering

Hummingbirds have been held in mystical awe since prehistoric times. A famous hummingbird icon is the Nazca image on the Peruvian desert near the town of Ica. The figure is over 750 feet long and can be seen only from the sky. Interestingly, a primitive culture created the figure on the desert thousands of years ago before the people could have viewed their giant hummingbird from the air. Early Americans believed that hummingbirds held the power over life and death, and that each night the birds die, only to come back to life in the morning. They were right, in a way. The hummingbird goes into a state of torpor at night, similar to hibernation. Its metabolism processes slow down until the morning sun warms it up and it "comes back to life."

Few birds command such intrigue as the hummingbird. The tiny bird's ability to hover as it takes nectar from a flower puts it in a class by itself. Like a Lilliputian helicopter, it floats forward, backward, up, and down with seeming ease. However, this effort takes an amazing amount of energy considering the size of the bird. A hummingbird might consume one-and-a-half to three times its weight in nectar each day to meet its energy demands. A human would need to eat 3,500 pancakes in one day to consume as much sugar as a hummingbird requires for flight.

The wings of a hummingbird are unlike those of other birds. Compared to other bird wings, the shoulder and forearm bones are greatly shortened. The portion of the wing that generates flight—both lift and thrust—is the hand portion consisting of the primaries. The hummingbird has only six secondary feathers, which are greatly reduced in size, comparatively.

The hummingbird is like a little helicopter. The wings rotate forward and backward in a figure-eight pattern. However, a helicopter has a propeller that revolves in a continuing 360-degree circle. That is beyond the ability of the hummingbird's wings. Instead, the hummingbird has some incredible three-dimensional abilities for flight in all directions that would make the best helicopter pilot envious. Each wing serves as half a propeller that rotates forward and backward in an approximately 180-degree arc. In reality, the wings function more like boat oars than helicopter propellers.

In level hovering flight, which a hummingbird typically uses when feeding at a flower, the wings move forward and backward in a horizontal plane. During the forward power stroke, the wing moves forward and slightly downward, with the front edge of the wing canted upward so that the upward angle of attack generates lift. The lift is generated by the angle of attack of the wings rather than by the secondary feathers as in other birds.

Once the wings are both straight forward, the shoulder bones rotate backward in their sockets so that the wings are upside down. They then thrust backward

Hovering

Opposite: Stage 1: A booted racquettail has its wings fully extended to the rear, ready for a power stroke forward.

and slightly downward with the leading edge higher than the trailing edge, so that the angle of attack again generates lift, and the wing is again acting as an oar. The angle of the wingbeat is directed slightly downward, making the complete path traced by the wing a figure-eight pattern. The rate of the wingbeats is extremely rapid. Most hummingbirds beat their wings at least twenty-five times per second. The ruby-throated hummingbird may exceed fifty strokes per second. The tiny bee hummingbird from Cuba is reported to beat its wings two hundred times per second.

When a hummingbird flies forward, the plane of the figure eight is tilted forward and slightly downward, directing the downwash of the wings backward. To fly backward, the plane of the figure eight is tilted upward and slightly backward, directing the downwash

forward. The hummingbird is the only bird that can fly backward. By canting its wings, it can fly sideways.

Some flower petals hang straight down, so a hummingbird may flip upside down and hover in a horizontal plane to feed. The hummingbird holds its upside-down body steady as it sips nectar from the flower while doing its unique version of an avian backstroke.

Opposite: Stage 2: A violet sabrewing is midway through the forward power stroke. The leading wing feathers have assumed the curving propeller profile that generates lift for hovering.

Above: Stage 3: At the completion of the forward power stroke, both wings of a male scintillant hummingbird are pointed forward.

Above: Stage 4: As a ruby-throated hummingbird begins the backstroke, the wings begin to turn upside down as the wing rotates in its socket. There is no venetian blind effect as on larger birds that use flapping flight.

Opposite: Stage 5: When the wings of a scintillant humming-bird are thrust backward, the upside-down wings create an angle of attack that generates lift. This continues until the wings are again fully extended to the rear.

Wind Hovering

Kestrels and terns exhibit a phenomenal type of flight called wind hovering that can look like levitation. Their open habitats offer few high perches to use for hunting prey. Instead, kestrels wind hover, facing into the wind, as they watch for the movement of grasshoppers or mice. Terns, ospreys, and kingfishers wind hover over water as they watch for minnows, small fish, and other aquatic prey.

To wind hover, a bird shifts its body posture from the typical horizontal flying position to nearly vertical. The wings are upraised, so the secondaries do not provide lift. The primary feathers, by acting as recipro-cating propellers, provide the lift needed to stay in place. With the tail outspread, the bird moves its primaries forward and backward. It thrusts the primaries forward to create enough lift to stay in position, then twists them backward in their sockets before beginning a backward stroke. The primaries open up like the blades of a venetian blind, minimizing air resistance. The slightly upturned plane of the upside-down primary feathers generates a small amount of supplemental lift. Once the primaries are back in their original position to the rear, they twist forward and overlap in preparation for another forward power stroke.

The wind hovering performed by the common tern is different from the hovering performed by the hummingbird. In this photo, the "arm" portions of the tern's wings are directed upward as the primaries are flapped forward and backward. This form of wind hovering provides no forward thrust but produces just enough lift to remain stationary and allow the bird to watch for prey below.

Boobies, gannets, and brown pelicans are marine waterbirds that hunt for small fish near the ocean's surface. The blue-footed booby flies thirty to forty feet above the ocean looking for food. Often, boobies hunt as a flock. When a school of fish is located, dozens of blue-footed boobies engage in a feeding frenzy that involves them all diving, surfacing, and taking off to dive again. Brown pelicans, gannets, and boobies are too heavy to wind hover the way a tern or kestrel does. Instead, they pause momentarily by flying into the wind and raising the front of their bodies to create a greater angle of attack. This position allows them to remain motionless for a short time while they scan the sea below and then bank and dive. To begin a dive, a blue-footed booby folds its primaries back. As it approaches the water's surface, the wings fold farther back to the extent that they point almost straight backward and then meet at the wingtips. The bird looks like an avian arrowhead as it plunges headfirst into the ocean. The booby may dive as deeply as thirty feet or more. Then it turns upward and captures fish during its return to the surface. Upon surfacing, the booby flaps its wings and runs along the surface of the water to take off. Like the booby, the brown pelican also spots fish while in flight, banks sharply, and dives. However, it captures fish in its bill as it enters the water.

PATTERING

Some storm petrels practice an unusual pattern of flight called pattering. The Elliott's storm petrel, which lives in the vicinity of the Galápagos Islands, flies in a slow, fluttering manner that allows it to splash its feet lightly on the water's surface. Pattering is a type of hovering in which the storm petrel's feet are suspended in the water and act as the tail of a kite. This stabilizes the flight and allows the petrel to travel at a very slow speed as it scans for invertebrates at the water's surface.

The Australasian gannet of New Zealand is not able to wind hover for extended periods the way terns and kestrels do, because it has high wing loading. However, the hunting gannet can pause above the sea and momentarily look for fish.

Near the Galápagos Islands, the Elliott's storm petrel carries out a unique fluttering flight as it tiptoes, or patters, across the water. This performance appears to help locate small prey invertebrates.

THE ART OF *Landing*

Considering the diverse manner in which birds fly—sometimes at speeds of two hundred miles per hour and in all manner of habitats and weather conditions—landing can be a suspense-filled maneuver. It involves intricate, dramatic moments. Depending on the age and experience of the bird, landing can reveal a range of technical skills, from comedic awkwardness to the elegance of an aerial ballet.

Many movements are necessary to carry out a successful landing. A bird has more control when it lands into the wind. It will usually circle a potential landing spot so it can face the wind as it lands. Then it selects a landing perch or surface where the chance of injury or accident is minimized. The bird also scans the landing area for potential predators.

A great blue heron executes a graceful landing, the final scene in an avian ballet. It is a challenge for a bird to decelerate from full flight and land at a chosen site without injury. *FloridaStock, Shutterstock*

141

For each species, there is a reason behind the technique that it uses in landing. Some birds, including hummingbirds, songbirds, and even brown pelicans and great egrets, must be able to make pinpoint landings in treetop branches, which requires great agility in landing technique. Albatrosses need the wide open spaces of the sea and bare sea islands to negotiate landing with great outspread wings and short tails that provide little maneuverability at the point of touchdown.

WATERSKI TO LAND

Pelicans often approach a landing area in a formation like a squadron of bombers. As the flock nears its destination, the birds' wings and bodies raise up to present a greater angle of attack. This angle of attack increases the drag and slows the airspeed as the pelicans descend to the water. The tails are spread wide and directed downward to add more drag. As the pelicans become more upright, the alulas lift up to prevent the birds from stalling. The wings at this

As a Canada goose lands, the body is slightly upraised, the tail is outspread to create drag, and the feet are lowered. The wings are slightly upturned in a "flaps down" position. Many of these actions are comparable to those of a landing plane, such as the "flaps down" position of the airplane wing.

Above: The great white pelican of Kenya begins its landing process by using its wings as a parachute to decelerate. The wings, tail, and feet are all outspread to decrease airspeed as it approaches the water.

Right: Then the pelican increases the angle of attack to increase the drag and continues decelerating as the feet are extended forward. With outspread wings providing balance, the pelican waterskis along the water's surface to slow down and finally drops into the water.

When a landing trumpeter swan makes contact with the water, the body is upraised to a steep angle of attack. The ruffled back feathers indicate that the airstream over the back has broken off and the swan is stalling as it touches down. The huge webbed feet serve as waterskis to help slow the swan as it lands.

When a cape petrel lands, it does not pull its body into as upright a position as swans and pelicans. It lands with a higher forward speed and glides to a stop using its webbed feet as waterskis.

point are upraised and flap forward and backward in a wind-hovering manner as they approach touchdown. The secondary wing feathers point downward in a flaps-down position that further reduces the final approach speed. The feet extend forward with the webbed toes outspread, serving as avian waterskis as the pelicans touch down, glide forward, and then settle into the water.

Trumpeter swans also offer a magnificent spectacle as they maneuver their graceful bodies for a picture-perfect landing. A swan circles its potential landing area, usually in formation with its mate and young of the year. They circle until they are flying into the wind. They then raise the forward portion of their bodies to increase the angle of attack. Flapping the wings in broad descending arcs, they begin their descent as their angle of attack approaches a 50 degree angle. The alulas raise to prevent stalling, the tail lowers, and the huge black

webbed feet extend forward with toes spread wide to present more air resistance and slow airspeed. Trumpeter swans typically touch down with outstretched feet as if waterskiing. As the forward speed slows, the swan's breast settles into the water and the wings are held partially open as the great swan glides to a stop, still honking and checking on the location of its mate and young.

The cape petrel is a seabird that keeps the company of albatrosses, shearwaters, and giant petrels off the coast of New Zealand and southern South America. About the size of a small duck, it has relatively short wings for a seabird. The cape petrel approaches the water with wings extended upward and outward. It lowers the trailing edges of the primary feathers to create more drag and lose velocity. Concurrently, the feet extend forward with webbed toes outspread to offer air resistance before landing and to waterski as it touches down.

The white-capped albatross uses its outspread webbed feet to slow its descent as it lands.

GLIDE TO LAND

Albatrosses, shearwaters, loons, and a variety of other waterbirds must deal with high wing loading and a fast approach speed as they land. These birds raise their bodies to increase the angle of attack, lower and spread the tails, and (for some) lower their webbed feet so that the toes drag on the water to slow their speed as they touch down.

The albatross is master of the air and can spend days aloft. However, it must land on the water to feed on squid and other aquatic creatures. It often lands near fishing vessels to eat fish parts discarded in the ocean. Its wings are perfectly adapted for a soaring life at sea. However, albatrosses and shearwaters have high wing loading that does not allow for a slow, graceful landing process. To land, the wings and body are upraised to increase the angle of attack so increased drag will reduce the airspeed. In contrast to geese and swans, which extend their webbed feet forward to slow their airspeed, the albatross hangs its feet down and spreads its webbed toes. At the point of touchdown, its feet drag on the water, it extends its wings upward to create maximum drag, and it glides forward, sliding breast-first until it comes to a stop.

Albatrosses typically land on water, but a nesting albatross must manage to land on an open area of ground without injuring itself. Often, those landing scenarios are less than graceful.

The shearwater is neither as large as the albatross nor are its wings as long, but it has high wing loading and a fast approach when landing on water. It also lands with a breast-first landing in which the breast helps cushion the impact when touching down.

The sooty shearwater uses a breast-first landing style to glide to a stop.

Opposite top: As a common loon glides in for a high-speed landing, it raises its angle of attack to approximately 20 degrees and extends its feet downward with the webbed toes spread. The short tail is directed downward to increase drag.

Opposite bottom: The extended feet are the first part of the loon to touch down. The toes dragging on the water decrease the airspeed further. Notice that the tail is still extended downward to slow the loon.

Below: As the loon glides to a stop, it gives the appearance of a torpedo that has just landed. Because of its high wing loading, the loon must land at a relatively high speed in open water with enough runway space to glide to a stop.

The common loon also has high wing loading, meaning that its body weighs a lot considering the small surface area of the wings. It flies with a rapid wingbeat and is unable to significantly slow its flight before touchdown. To land, the body is slightly upraised so that the angle of attack reaches 8 to 10 degrees above horizontal. The wings are outspread and canted upward to increase the drag. The large webbed feet extend down so that the top of the feet offer additional air resistance.

As the loon approaches the water's surface, it drags its feet on the water to aid in decelerating. As it loses velocity, it maintains uplifted wings and glides forward in a chest-first posture, like a feathered torpedo. The feet of a loon are located so far back on the body that if it extended its feet forward to land, the bird would flip forward when it hit the water.

The lappet-faced vulture of Kenya appears to float from the sky with wings outspread like the sails of a hang glider. The feet are lowered like landing gears and help absorb the jarring effect of landing. The large tail is outspread and angled downward to aid in deceleration.

The lappet-faced vulture makes a graceful gliding descent as it lands on the plains of eastern Africa. The wings are characterized by lower wing loading, which permits a slower landing approach. With wings fully outspread, the tail lowers to increase drag and the feet lower like the landing gears of a plane. The vulture floats to earth with the slow grace of a hang glider. At that point, all references to grace and beauty disappear as the vulture begins pecking away at a dead wildebeest.

PULL UP, STALL, AND DESCEND

The egret presents a contradiction in form and function. How can such a large bird, sporting a four-foot wingspread and long legs, nest in treetops and land safely among a tangle of branches? The answer seems to be: with elegant grace.

Great egrets and snowy egrets have long, broad wings that have a low aspect ratio and provide considerable lift. The long legs provide extra reach for grasping branches and are not as awkward as they might appear. Egrets live in treetop colonies that may contain hundreds of egrets and herons. To land in such a crowded nesting area, the body is uplifted to an upright angle of attack that may approach 45 degrees above horizontal. At this point, the bird stalls

Above: A brown pelican lands in a treetop nesting colony in Costa Rica in a manner similar to that used by egrets and herons. The body is upraised and the wings and tail are outspread as the pelican stalls and gracefully descends to its perch.

Left: The large wings of the snowy egret serve as parachutes that allow the bird to descend gracefully to its treetop perch. The ruffled feathers on the backside of the wings give evidence that the bird has intentionally stalled prior to landing and that the stream of air over the back has been replaced by turbulence.

and aerodynamic forces cease to function—except gravity. With wings outspread, the egret gently descends onto the small branch it has selected as its perch. When a photographer captures this point of stalling on film, it appears as if the egret is magically levitating above the treetops.

Like the egret, the brown pelican nests in treetops. It inhabits coastal areas of the Atlantic, Caribbean, and Pacific south to Peru. Graceful in flight, the bird demonstrates exceptional agility as it dives for marine prey from the sky and alights among the branches of its nesting area. To land in its treetop nest, the brown pelican raises its body posture sharply, increasing the angle of attack and intentionally stalling. The huge outspread wings serve as parachutes as the pelican gently descends. The outspread tail, which has twice as many feathers as most other birds, provides an effective rudder for maneuvering and decelerating.

In contrast, the American white pelican does not demonstrate the aerial dexterity of the brown pelican. The white pelican feeds while swimming, not diving from the air, and it nests on the ground of islands and peninsulas of larger lakes of the upper Midwest, Great Plains, and central Canada.

Swallows, martins, tanagers, flycatchers, crows, and smaller songbirds reflect life in the fast lane when they approach a perch by rising to a nearly upright position. The angle of attack results in intentional stalling in which the forward momentum of the bird gives it just enough forward thrust to grasp the perch and land. As it grasps a twig or other elevated perch, the forward momentum then carries the bird to a horizontal perching position.

Right: To land, a purple martin rises to a nearly upright posture and extends its feet to grasp the wire. At this point, the wings no longer provide lift. The forward momentum of the bird carries it to the wire and forward to a horizontal posture.

Some birds appear less than graceful when they land. This Chihuahuan raven is flying to a treetop perch in Arizona. When it reaches the intended perch, it will pull up to a nearly upright posture and grasp the branch, and its forward momentum will carry it forward to a horizontal resting position.

Above: The bay-headed tanager of Trinidad arrives at its landing site in an upright position and grasps the perch. Forward momentum causes its body to pivot forward and assume a horizontal perching position.

Left: When a songbird lands on the ground or, in this case, on the snow, it does not rise up to the extent of a bird landing on an aerial perch. This common redpoll of Russia has raised its body slightly as it spread its wings to alight. The tail is twisted to the side to help guide the bird to its intended landing spot.

WINGS UP, HOVER DOWN

To land, Canada geese and mallards circle into the wind, raise the body upward, and drop the feet downward. The landing bird raises its angle of attack to 20 degrees or more during the final descent. Then it undergoes a change from a strong forward flapping flight to an upright posture. The alulas are raised to avoid stalling. Then it raises the wings above the back and flaps the wings in a forward and backward manner, much as a kestrel does when wind hovering. The tail is spread wide to provide maximum braking.

When both wings are held above the back, the secondaries cease to provide lift. The primaries repeatedly flip forward and backward in a horizontal plane that provides just enough lift to allow the bird to descend slowly. The primaries serve as reciprocating propellers. When the wings are flipped backward, the primary feathers are in an upside-down position so that the air passes through the primaries like an open venetian blind.

This manner of landing allows the Canada goose or mallard to descend to the water with a more vertical approach than is possible with the trumpeter swan, which must approach the water in a horizontal flight path at a higher speed. A vertical descent can be especially useful when a bird is trying to land in an open spot in a marsh or field among hundreds or thousands of other waterfowl.

HOVER TO LAND

The widespread ruby-throated hummingbird provides a striking contrast in landing technique. While approaching a perch, it changes the plane of its hovering pattern. When the plane is level, the hummingbird can hover in place. However, by tilting the plane of its figure-eight wingbeat down toward the front, a backwash is created that propels the hummingbird forward and allows it to gently descend to its desired perch.

A ruby-throated hummingbird hovers as it approaches its perch. The body is upraised and the tail lowered and spread to slow its approach.

Opposite: The mallard is adapted to hovering downward to land. The body raises to a nearly vertical posture, and the tail spreads out to stop forward progress so the duck can drop in among other ducks. The wings raise and the primaries beat forward and backward to generate a moderate amount of lift as the duck descends.

Afterword

In the last pages of his classic book *Art of Birds*, Pablo Neruda wrote of his lifelong passion for birds and of his fascination for birds in flight. Following are the closing lines from his epilogue:

> *A people's poet,*
> *provincial and birder,*
> *I've wandered the world in search of life:*
> *bird by bird I've come to know the earth:*
> *discovered where the fire flames aloft:*
> *the expenditure of energy*
> *and my disinterestedness were rewarded,*
> *even though no one paid me for it,*
> *because I received those wings in my soul*
> *and immobility never held me down.*

You don't need to be a Nobel Prize–winning poet to appreciate the beauty of birds in flight. If you take time to watch the birds, you will begin to discover their remarkable adaptations for survival, migration, and flight that make them masters of the sky, and your life will be enriched beyond your wildest hopes. In the preceding chapters, I have shared some of my favorite photos of birds in flight and provided a general introduction to the aerodynamic principles of avian flight. Yet, even as I attempt to understand those principles and share them with you, I am still awed by the elemental beauty and magic of birds in flight that transcend the physical laws of nature.

Take the time to explore state and national wildlife refuges, parks, and forests. You will discover the beauty of birds in flight and the value of preserving these wild places for both wildlife and humans.
Terrance Emerson, Shutterstock

BIBLIOGRAPHY

Attenborough, David. *The Life of Birds*. London: British Broadcasting Corporation, 1998.

Aymar, Gordon C. *Bird Flight*. Garden City, NY: Garden City Publishing, 1938.

Brooke, Michael, and Tim Birkhead, eds. *The Cambridge Encyclopedia of Ornithology*. Cambridge, England: Cambridge University Press, 1991.

Burton, Robert. *Bird Flight*. New York: Facts on File, 1990.

Dalton, Stephen. *The Miracle of Flight*. New York: McGraw-Hill, 1977.

Haining, Peter. *The Compleat Birdman: An Illustrated History of Man-Powered Flight*. New York: St. Martin's, 1976.

Koch, Ludwig, ed. *The Encyclopedia of British Birds*. London: Waverly Book Company, 1955.

Levin, Alan. "Flying Creatures May Help Create Aviation of the Future." *USA Today*, February 13, 2007: 4A.

Neruda, Pablo. *Art of Birds*. Barcelona: Lynx Edicions, 2002.

——. *Selected Odes of Pablo Neruda*. Berkeley: University of California Press, 1990.

Page, Jake, and Eugene S. Morton. *Lords of the Air: The Smithsonian Book of Birds*. London: Orion Books, 1989.

Proctor, Noble, and Patrick J. Lynch. *Manual of Ornithology*. New Haven, CT: Yale University Press, 1993.

Rüppell, Georg. *Bird Flight*. New York: Van Nostrand Reinhold, 1975.

Safina, Carl. "On the Wings of the Albatross." *National Geographic* 212 (2007): 86–113.

Stillson, Blanche. *Wings: Insects, Birds, Men*. New York: Bobbs-Merrill, 1954.

Storer, John H. *The Flight of Birds*. Bulletin No. 28. Bloomfield Hills, MI: Cranbrook Institute of Science, 1948.

Terres, John K. *The Audubon Society Encyclopedia of North American Birds*. New York: Knopf, 1980.

Van Tyne, Josselyn, and Andrew J. Berger. *Fundamentals of Ornithology*. New York: John Wiley and Sons, 1961.

Weimerskirch, Henri. "Wherever the Wind May Blow." *Natural History* 113 (2004): 40–45.

The long-tailed sylph hummingbird provides an elfin vision of avian beauty while visiting a feeder in the Venezuelan rainforest.

INDEX

Author Carrol Henderson and his wife, Ethelle. *Photograph by Ray and Marlene Simon*

ABOUT THE AUTHOR/PHOTOGRAPHER

Carrol L. Henderson is a wildlife biologist with the Minnesota Department of Natural Resources. He has served as the DNR's Nongame Wildlife Program supervisor since the program's inception in 1977. During that time, he has developed a statewide wildlife conservation program that has been one of the most successful programs of its kind in the nation. Species he has helped reintroduce include eastern bluebirds, peregrine falcons, bald eagles, trumpeter swans, and river otters.

Henderson is the author and photographer of *Oology and Ralph's Talking Eggs: Bird Conservation Comes Out of Its Shell* and *Field Guide to the Wildlife of Costa Rica*. He has also written or co-authored five books for the Minnesota Department of Natural Resources:

Woodworking for Wildlife, Landscaping for Wildlife, Wild about Birds, Traveler's Guide to Wildlife in Minnesota, and *Lakescaping for Wildlife and Water Quality.*

Henderson and his wife, Ethelle, have been leading international birdwatching trips since 1987. They have traveled throughout Central and South America, New Zealand, Kenya, and Tanzania. Henderson is an enthusiastic nature photographer who has photographed over a thousand species of birds. He has a collection of over 70,000 wildlife and nature photos from around the world. His photos have appeared in the *New York Times, Audubon, Wild Bird, Birder's World,* and the *World Book Yearbook of Science.* Henderson and his wife live in Blaine, Minnesota.